OpenSCAD

Basics Tutorials

Tutorial Books

For resource files contact us at
openscadtut@gmail.com

Table of Contents

Getting Started with OpenSCAD

Introduction to OpenSCAD

OpenSCAD is a solid 3D modeler that allows you to create parametric models using its scripting language. You can create models by transforming and combining simple shapes. In OpenSCAD, everything is controlled by the parameters. For example, if you want to change the position of the hole shown in the figure, you need to change the **X** distance value of the **translate ()** function.

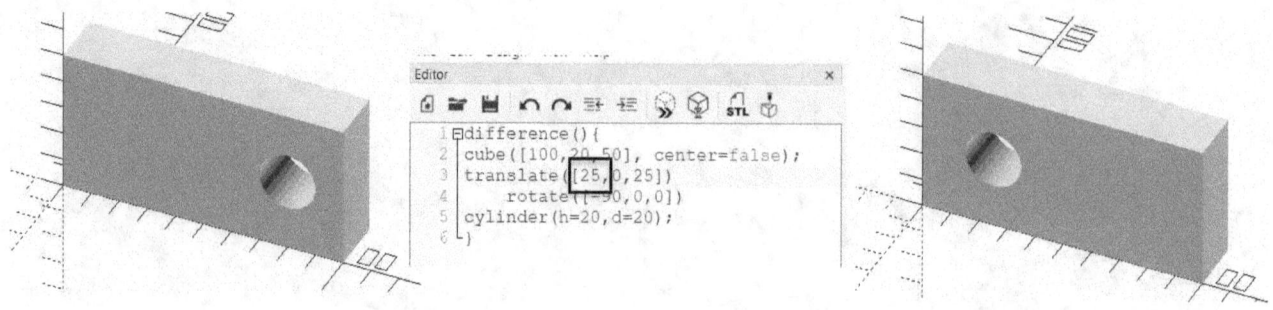

The parameters that you set up allow you to have control over the design intent. The design intent describes the way your 3D model will behave when you modify the dimensions of the model. For example, if you want to position the hole at the center of the block, one way is to manually change the **x** distance value of the **translate ()** function. However, when you change the size of the cube, the hole will not be at the center.

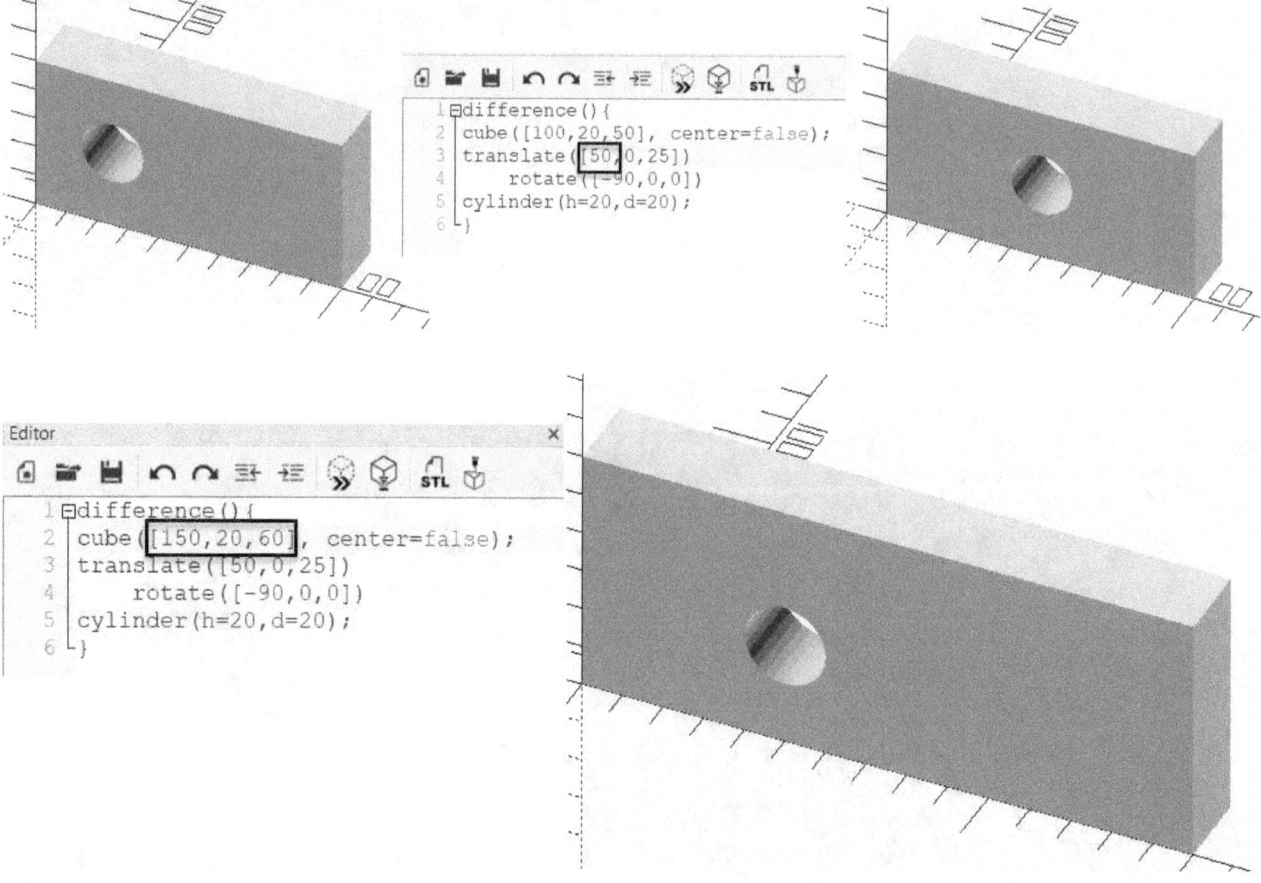

You can make the hole to be at the center, even if the size of the cube changes. To do this, you need to create the parameters of the model at the beginning of the script, as shown. Next, use the parameters in place of the values in the script. For example, change the **x** and **z** values of the **translate ()** function to the **cube_width/2** and **cube_height/2**, respectively.

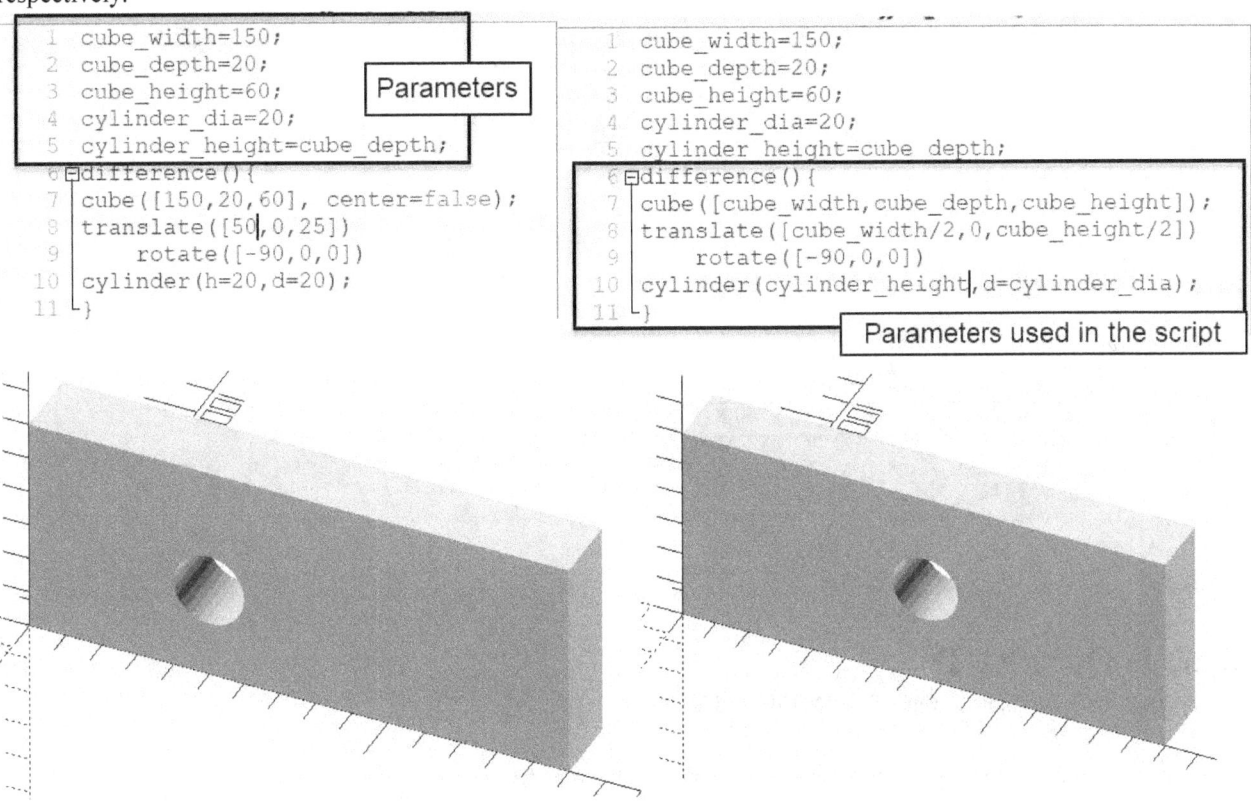

Now, even if you change the size of the cube, the hole will always remain at the center.

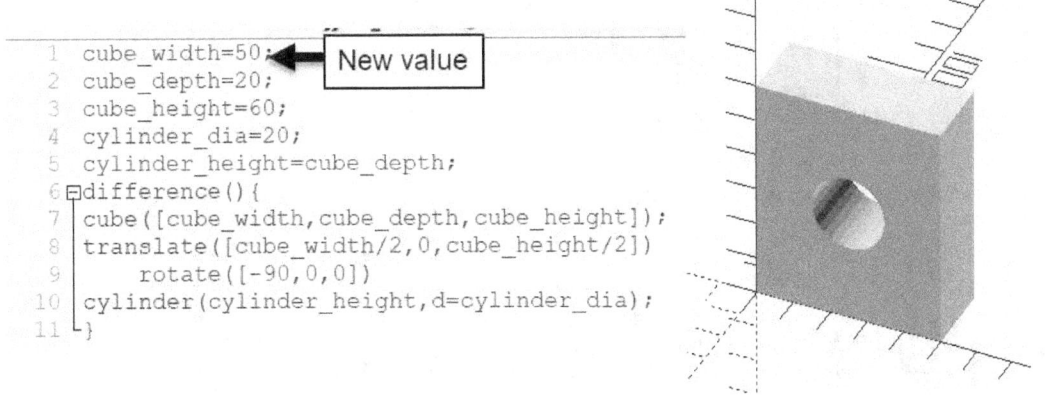

User Interface

The following image shows the **OpenSCAD** window.

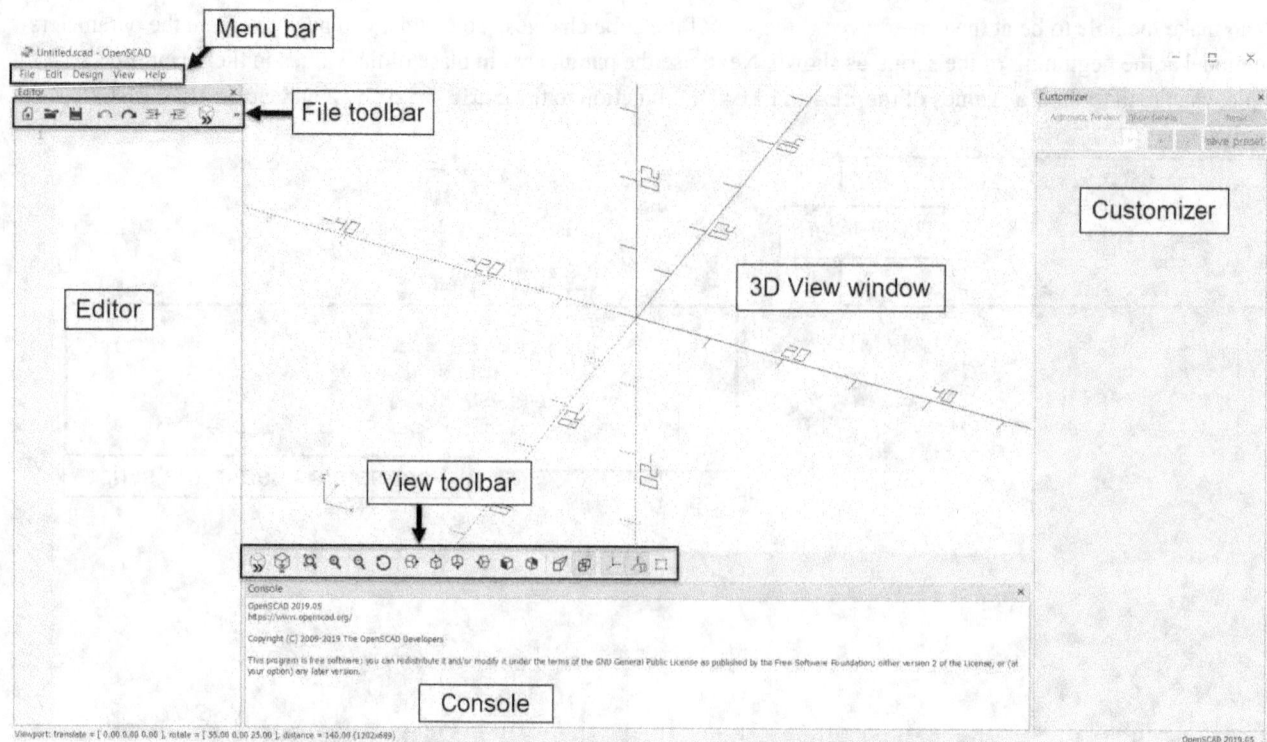

Toolbars in OpenSCAD

There are two toolbars available in OpenSCAD: **File** and **View**.

The **File** toolbar has options to create files, open, save, render, and export files.

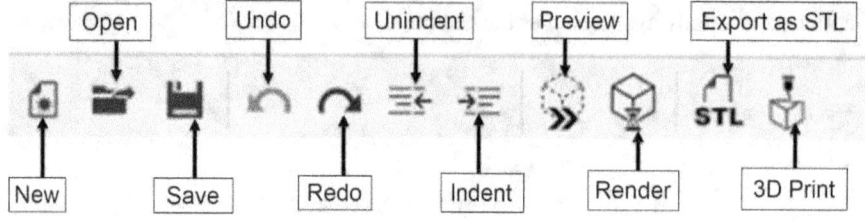

The **View** toolbar has the icons to modify the display of the model.

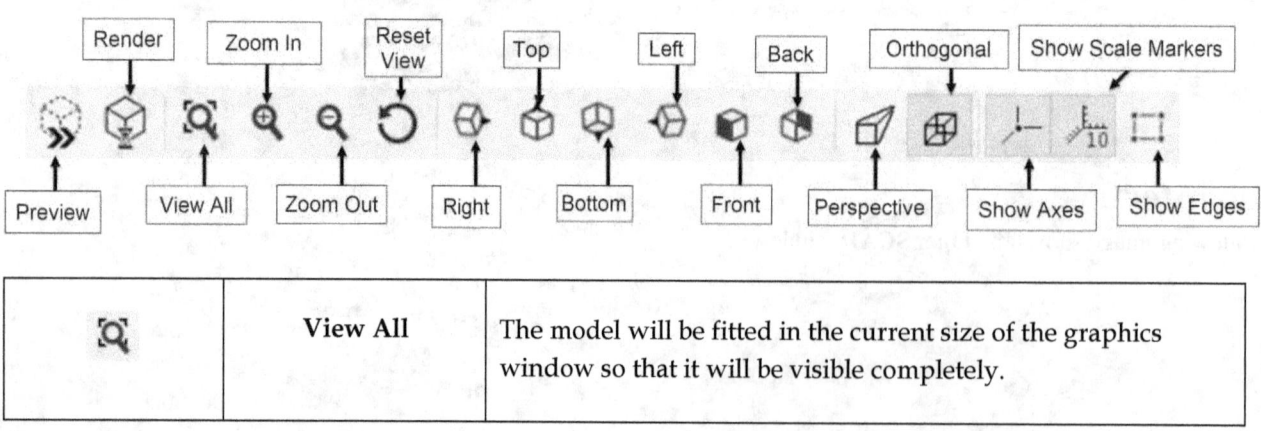

![View All icon]	**View All**	The model will be fitted in the current size of the graphics window so that it will be visible completely.

		Rotate	Press and hold the left mouse button. Next, drag the pointer to rotate the model view.
		Zoom In	Click this icon to the zoom in to the model.
		Zoom Out	Click this icon to zoom out of the model.
		Perspective	This icon allows you to switch to the perspective projection of the model.
		Orthogonal	This icon allows you to switch to the parallel projection of the model.
		Reset View	This icon allows you to reset the view orientation to Isometric.
		Right	This icon allows you to change the view orientation to right.
		Top	This icon allows you to change the view orientation to top.
		Bottom	Click this icon to change the view orientation to the bottom.
		Left	Click this icon to change the view orientation to the left.
		Front	Click this icon to change the view orientation to the front.
		Back	Click this icon to change the view orientation to back.

Menu Bar

The menu bar is located at the top left corner of the window. It consists of **File, Edit, Design, View,** and **Help** menus.

File   Edit   Design   View   Help

Editor window

The **Editor** window is located on the left side. It is used to enter the script to create the 2D and 3D objects.

11

3D View window

The 3D View window is located on the right side of the **Editor** window. It displays the 3D Model generated from the script that you enter in the **Editor** window.

OpenSCAD Help

OpenSCAD offers you the help system that goes beyond basic command definition. You can access OpenSCAD help by clicking on the **Help menu** icon on the right side of the window.

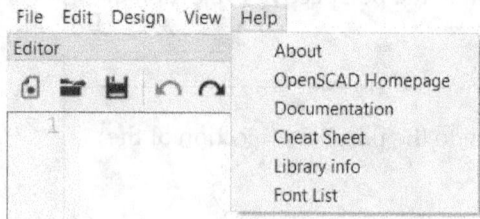

Tutorial 1

In this example, you create the part shown below.

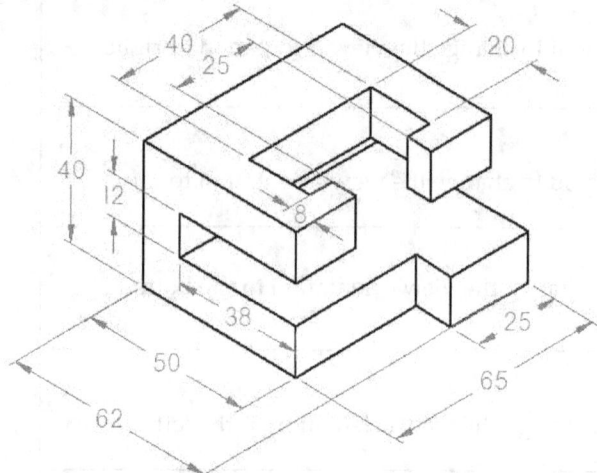

Creating a New Document

1. Click the **Windows** icon located at the bottom left corner of the desktop. Next, click **O > OpenSCAD**.

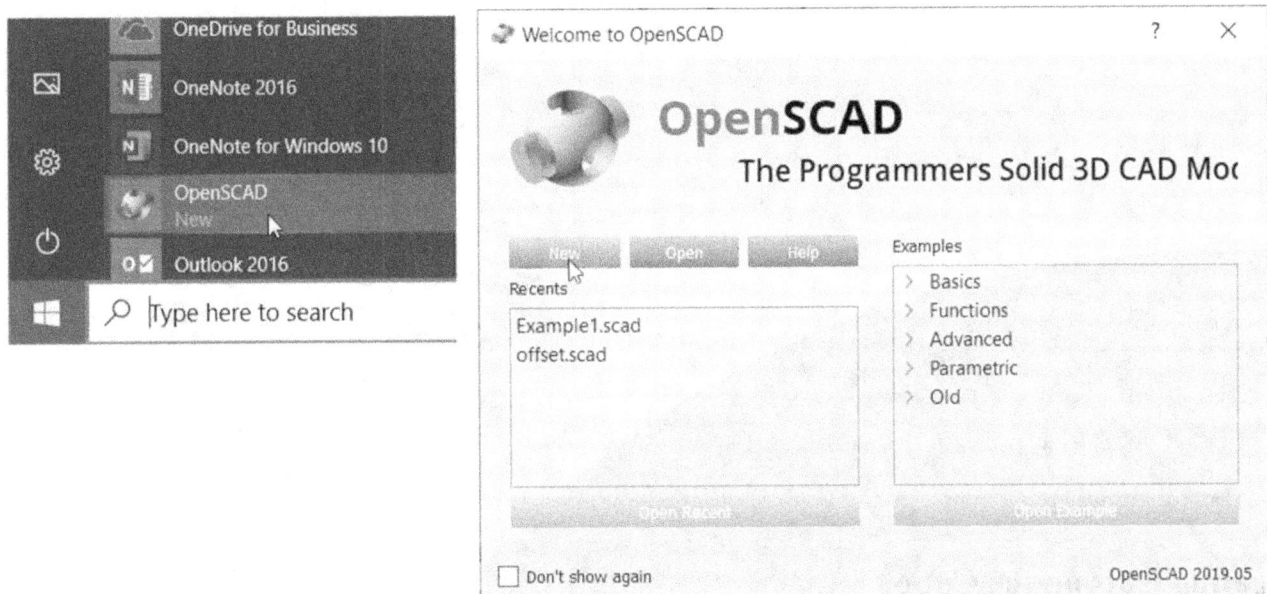

2. Click the **New** button on the **Welcome to OpenSCAD** dialog.

Creating a 3D Cube

1. On the menu bar, click **Help > Cheat Sheet**.
2. On the **Cheet Sheet** page, select the **cube([width,depth,height], center)** from the **3D** section.
3. Next, right-click and select **Copy**.

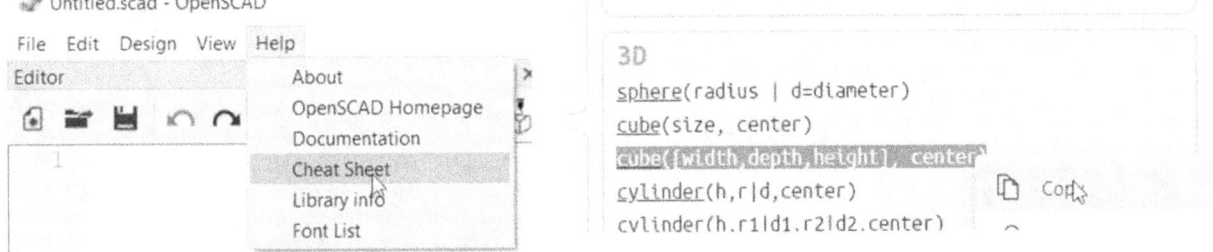

4. Switch to the OpenSCAD window. Next, right-click in the **Editor** and select **Paste**.
5. Change the width, depth, and height values to 50, 65, and 40, respectively.
6. Click next to **center** and type =false; the cube will not be created at the center of the **3D View** window.
7. Add semicolon ';' at the end of the script.

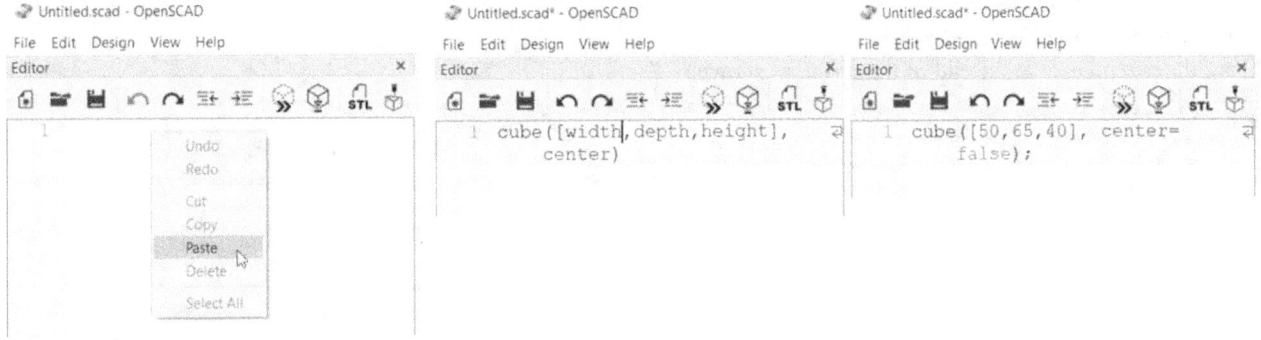

8. Click the **Render** icon on the **File** toolbar; the model is rendered.
9. Click the **View all** icon on the **View** toolbar; the model is fitted in the **3D View** window.

13

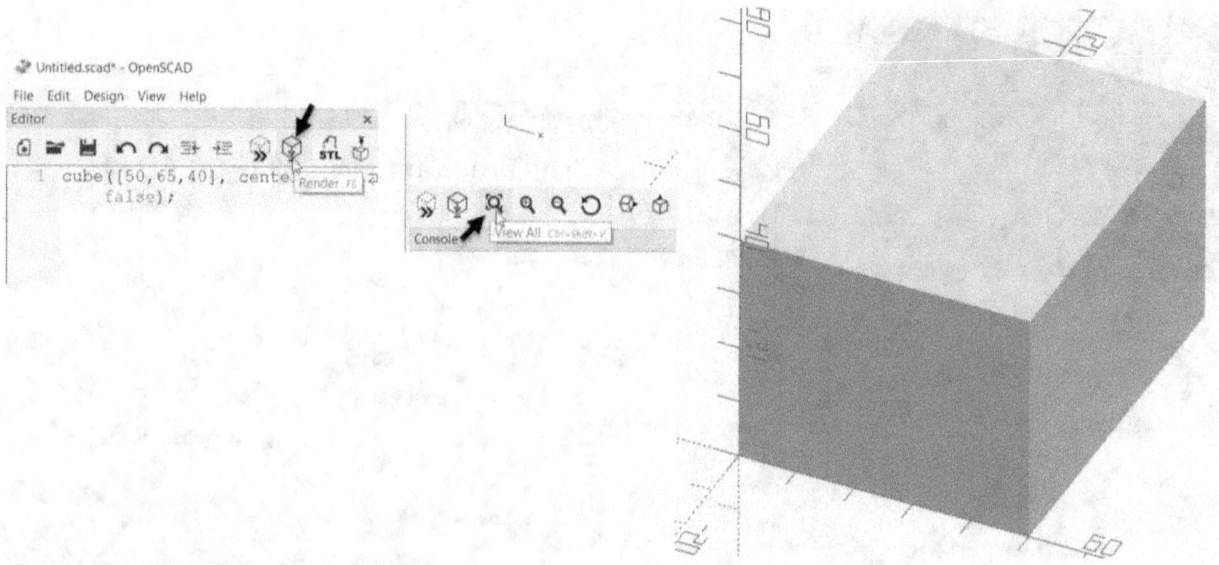

Creating Cuts on the Cube

Now, you need to create a cut on the cube using another cube. To do this, you need to create a cube, and then translate it to the required position. Next, subtract the cube from the first cube using the **difference ()** Boolean operation.

1. Click next to the semicolon of the cube function and Press ENTER.
2. Type **translate ([12,0,14])** and then press ENTER.
3. Type **cube ([38,65,12], center=false);** and then press ENTER. A cube of 38 width, 65 depth, and 12 height will be created. Also, it will be translated to the distance of 12 in the X direction, 0 in the Y direction, and 14 in the Z direction.

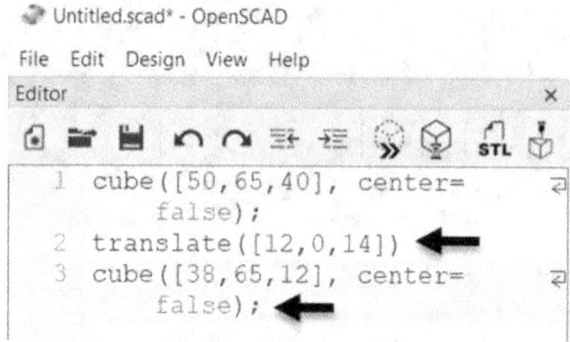

4. Click before the first cube.
5. Type **difference(){** and press ENTER.
6. Close the parentheses at the end of the script, as shown.

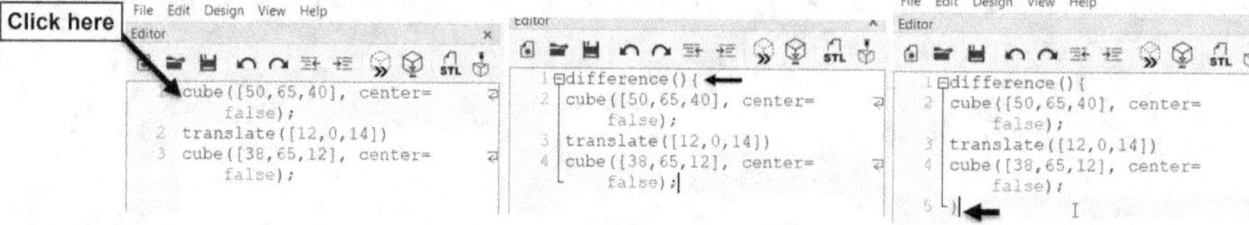

7. Click the **Render** icon on the **View** toolbar; the model is rendered.

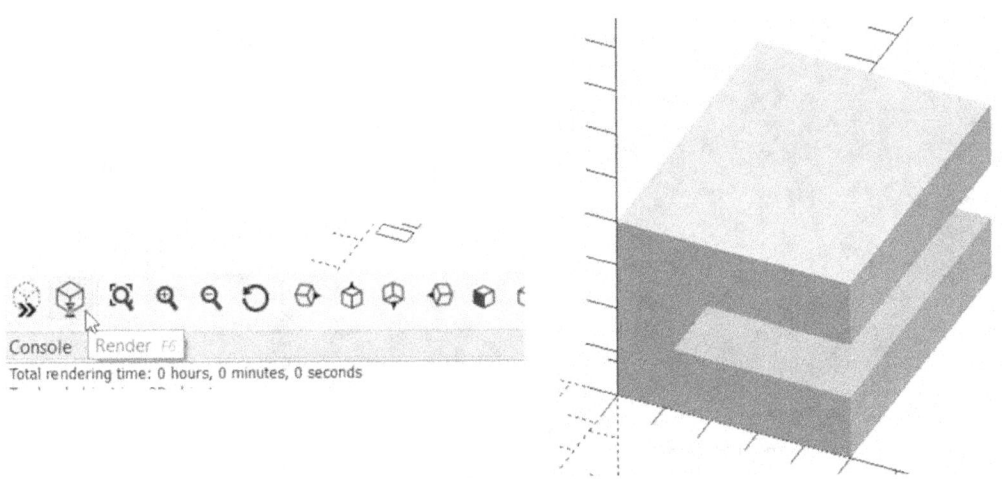

8. Select the translate and **cube** function. Next, right-click and select **Copy**.
9. Click next to the semicolon of the last **cube** function and press ENTER.
10. Right-click and select **Paste**.

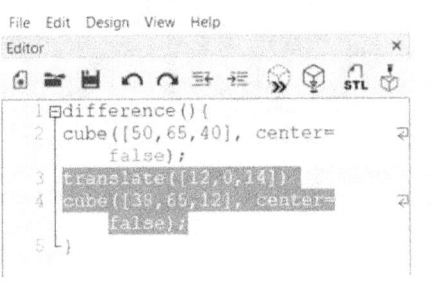

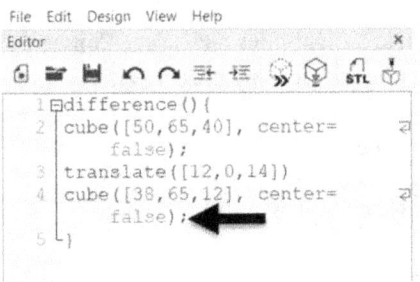

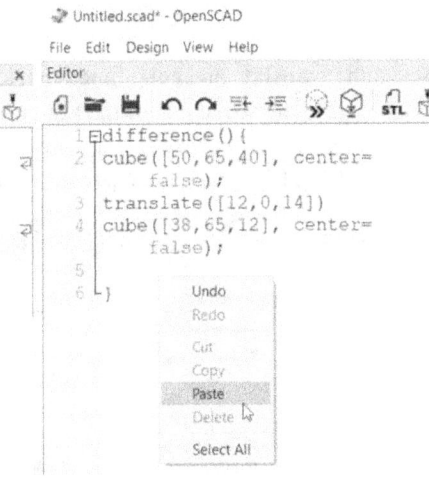

11. Change the values of the **translate** and **cube** functions, as shown.

12. Click the **Render** icon on the **View** toolbar; the model is rendered.

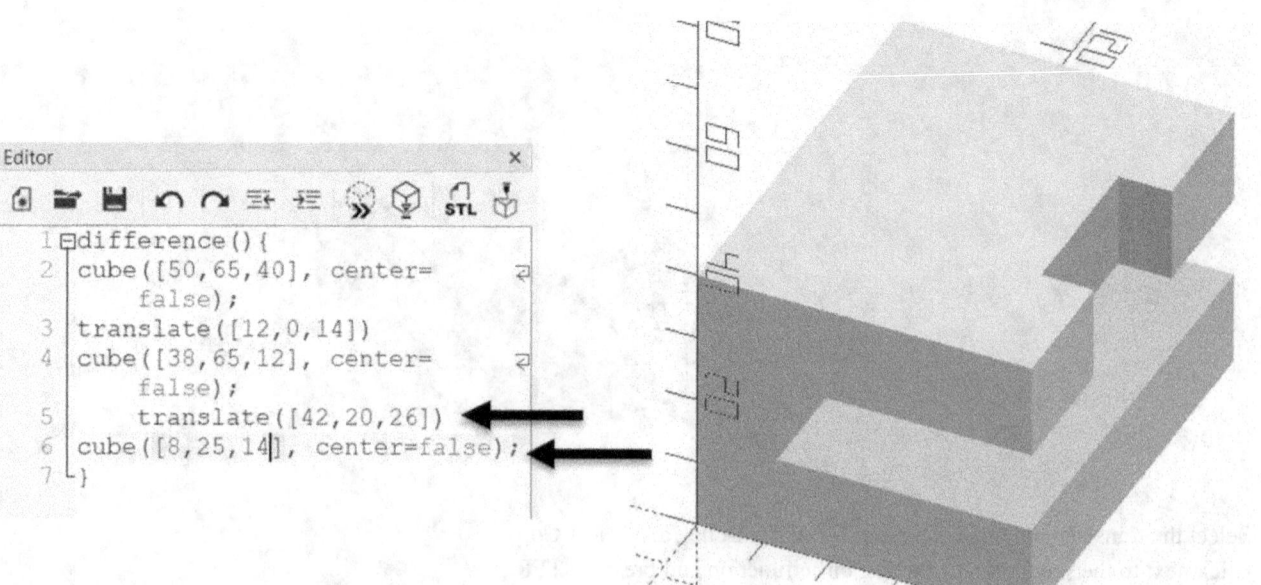

13. Select the **translate** and **cube** function. Next, right-click and select **Copy**.
14. Click next to the semicolon of the last **cube** function and press ENTER.
15. Right-click and select **Paste**.
16. Change the values of the **translate** and **cube** functions, as shown.

17. Click the **Render** icon on the **View** toolbar; the model is rendered.

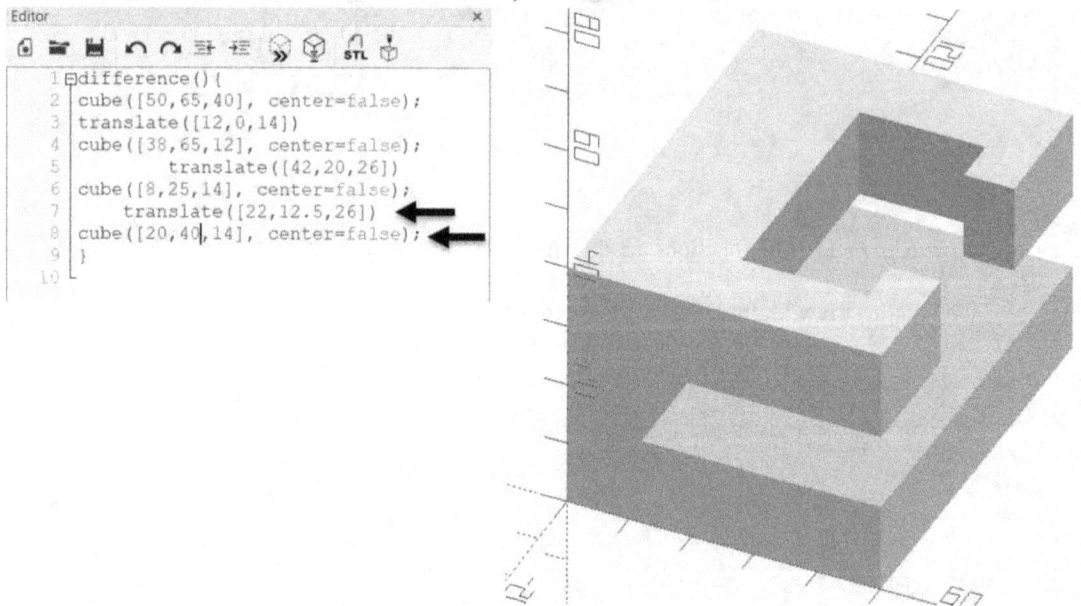

Adding a Cube to the model

Now, you will add a cube to the model using the cube, translate, and union functions.

1. Click at the beginning of the script and press ENTER.
2. Type **union(){**.
18. Click at the end of the script and press ENTER. Next, type **translate ([50,40,0])** and then press ENTER.

3. Type **cube ([12,25,14], center=false);** and then press ENTER.
4. Close the parentheses at the end of the script, as shown.

5. Click the **Render** icon on the **View** toolbar; the model is rendered.

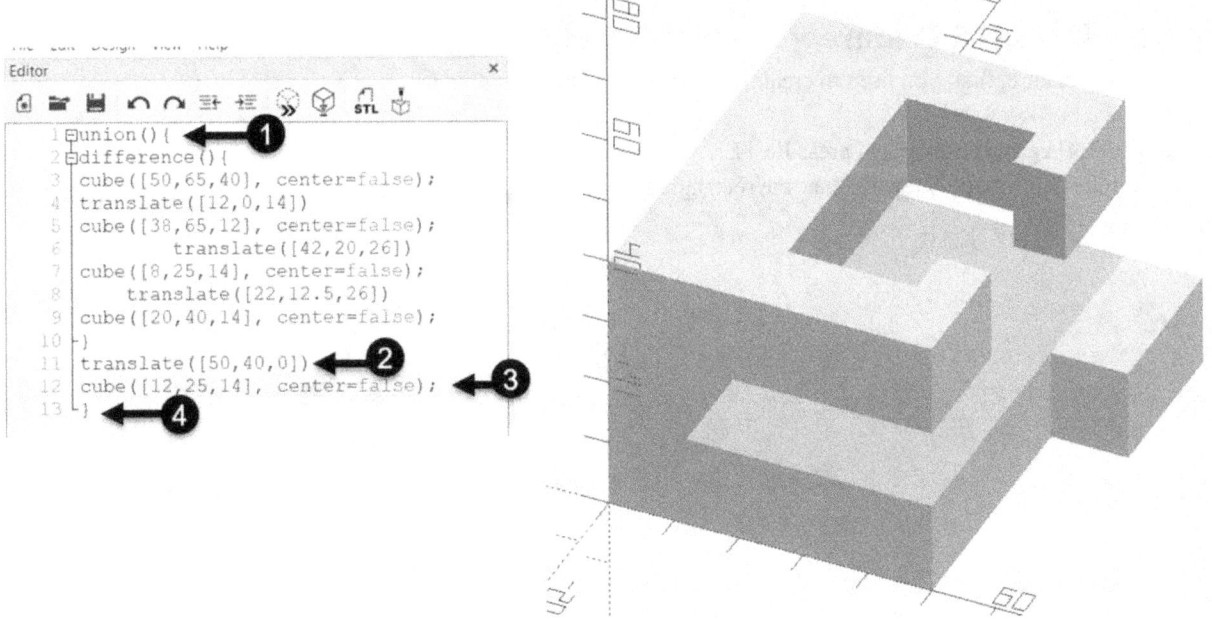

6. Click the **Save** icon on the **File** toolbar. Next, browse to the required location on your computer.
7. Type **Tutorial_1** in the **File name** box and click **Save**.
8. Click **File > Close** on the menu bar.

Tutorial 2

In this example, you create the part shown below.

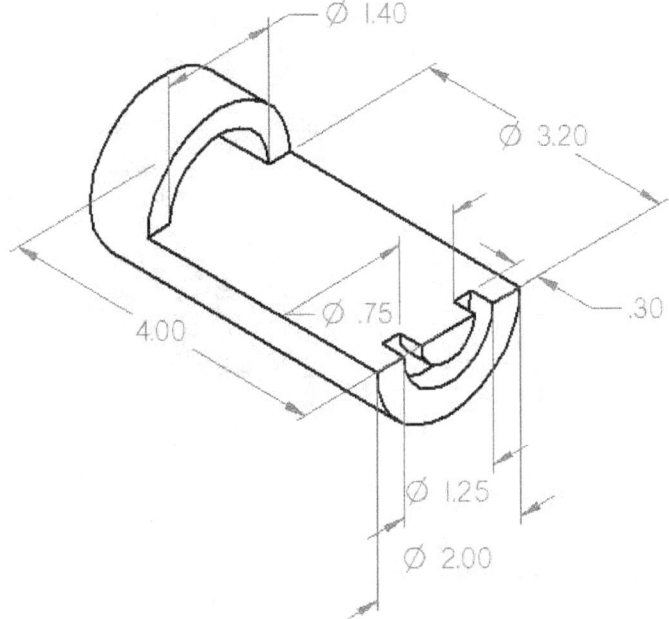

Creating a New document

1. Click the **Windows** icon located at the bottom left corner of the desktop. Next, click **O > OpenSCAD**.
2. Click the **New** button on the **Welcome to OpenSCAD** dialog.

Creating the Revolved Feature

To create a revolved feature, first, you need to create a rectangle, and then revolve it.

1. Type **square([1,4],center=false);** and press ENTER.
2. Click the **Render** icon on the **View** toolbar; the rectangle is created, as shown.

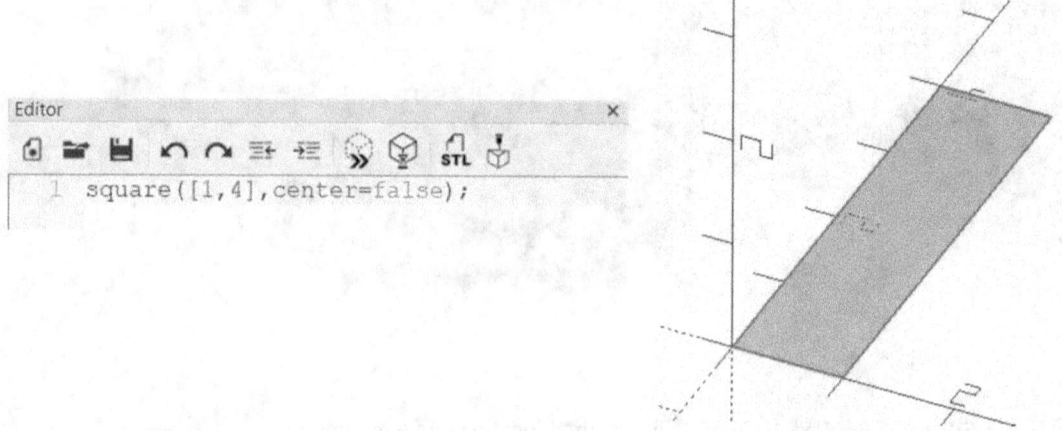

3. Click at the beginning of the **square** function. Next, type **rotate_extrude(angle=180)** and press ENTER.

4. Click the **Render** icon on the **File** toolbar; the square is rotate extruded by 180 degrees. Notice that the model is created in low resolution. You need to use the **$fn** function to increase the resolution of the model.

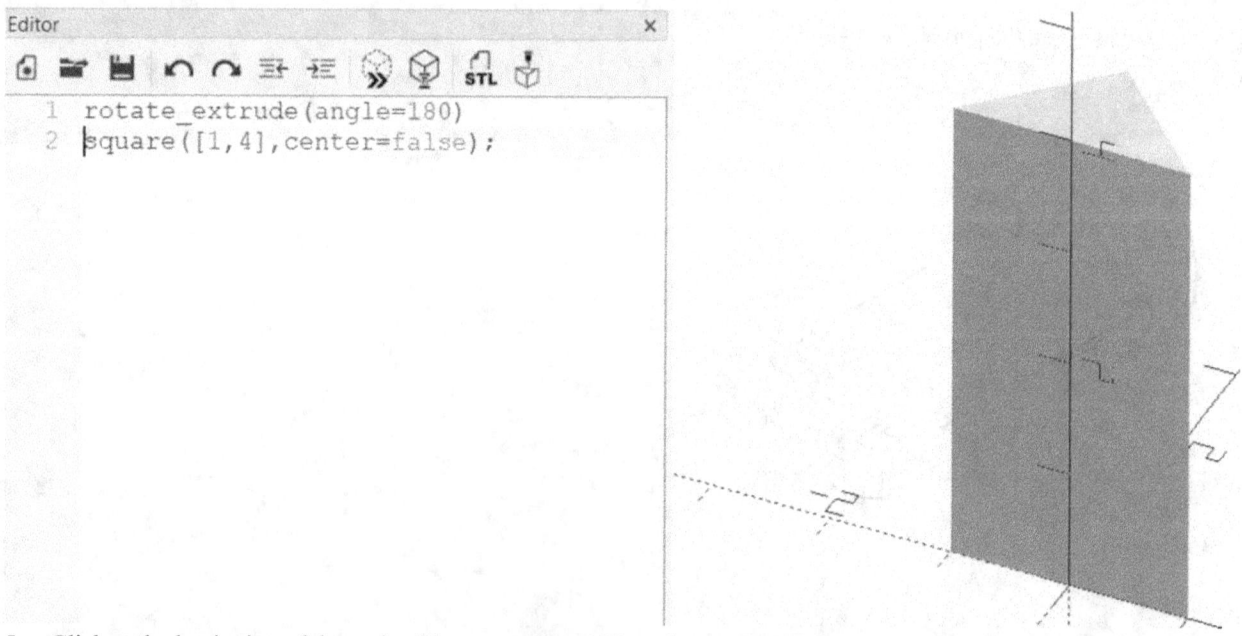

5. Click at the beginning of the script. Next, type **$fn=125;** and press ENTER.

6. Click the **Render** icon; the rotate extruded feature is smoothened.

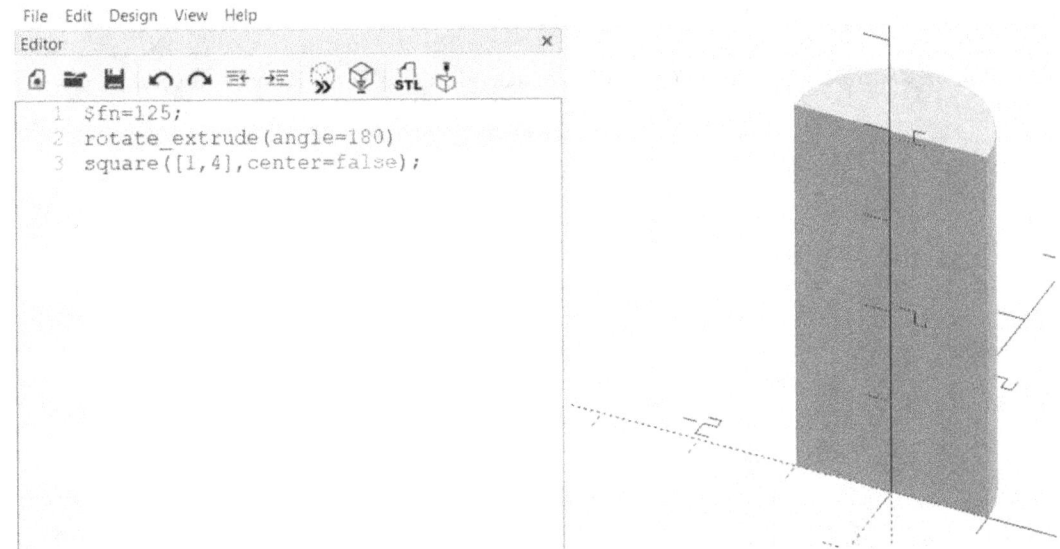

Creating a Revolved Cut feature

1. Click at the end of the script and press ENTER.
2. Type **square([0.375,0.3],center=false);** and press ENTER.

3. Click the **Render** icon; the rectangle is created. However, the rectangle is not visible.
4. To view the rectangle. Select the **rotate_extrude** and **square** functions above the newly created function. Next, click **Edit > Comment** on the menu bar, and then click the **Render** icon.

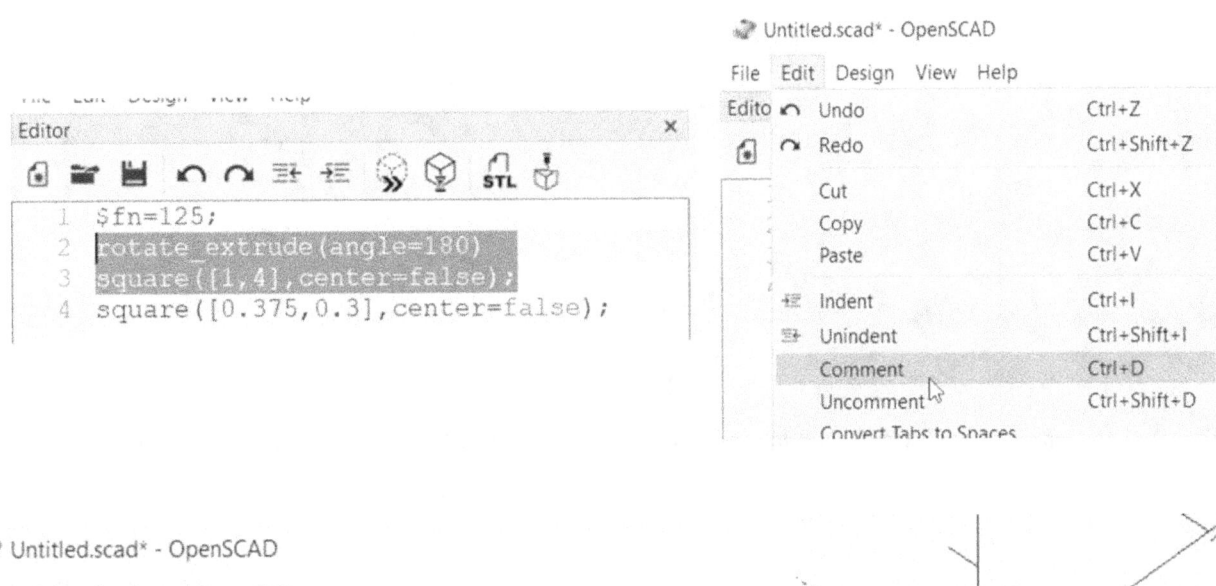

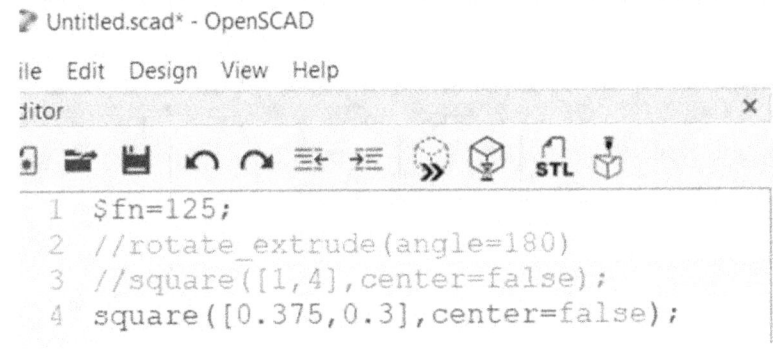

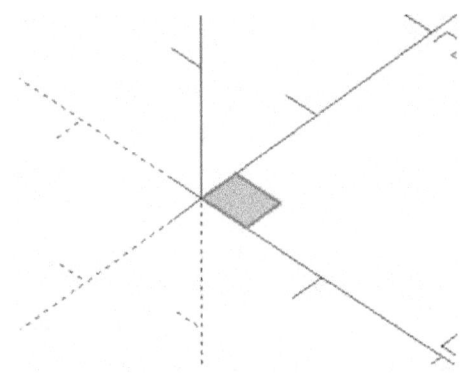

5.   Click before the **square** function and type **translate ([0.375,3.7,0])** and then press ENTER.

6.   Click the **Render** icon; the square is translated in up to **0.375** and **3.7** distances in the **X** and **Y** directions.

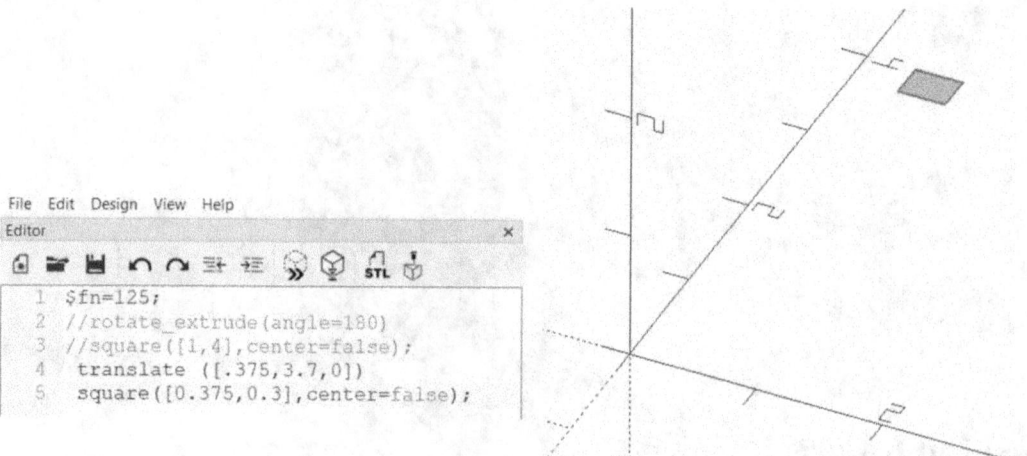

7.   Click at the beginning of the **translate** function. Next, type **rotate_extrude(angle=180)** and press ENTER.
8.   Click the **Render** icon on the **File** toolbar. The square is rotate extruded by 180 degrees

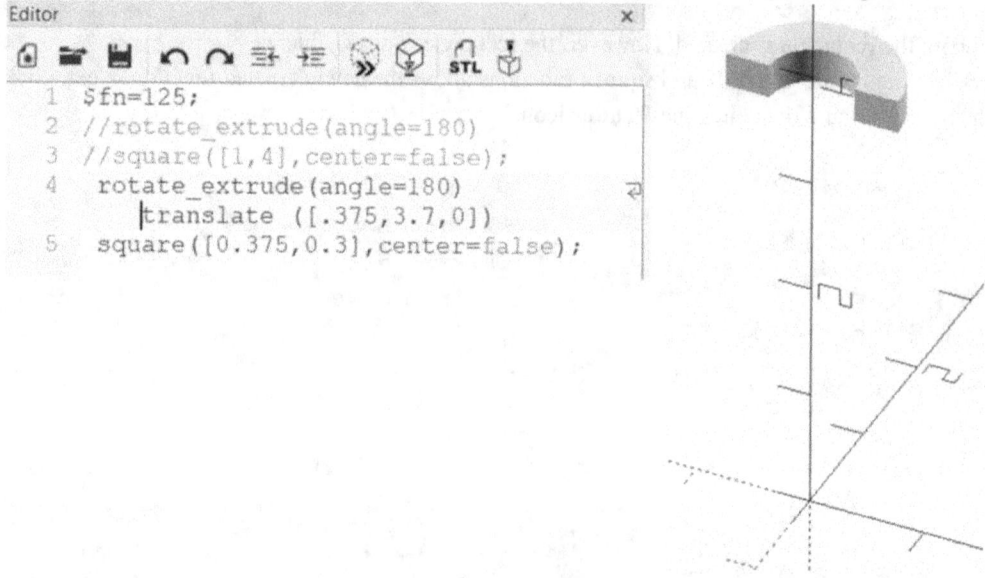

9.   Select the commented **rotate_extrude** and **square** functions. Next, click **Edit > Uncomment** on the menu bar.

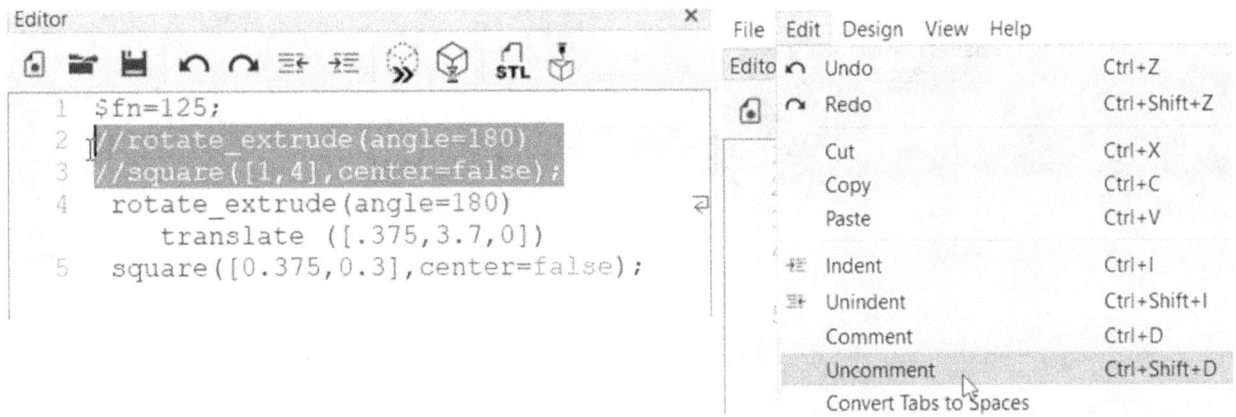

10. Click before the first **rotate_extrude** function.
11. Type **difference(){** and press ENTER.
12. Close the parentheses at the end of the script.
13. Click the **Render** icon on the **View toolbar**; the newly created rotate extruded feature is removed from the first one.

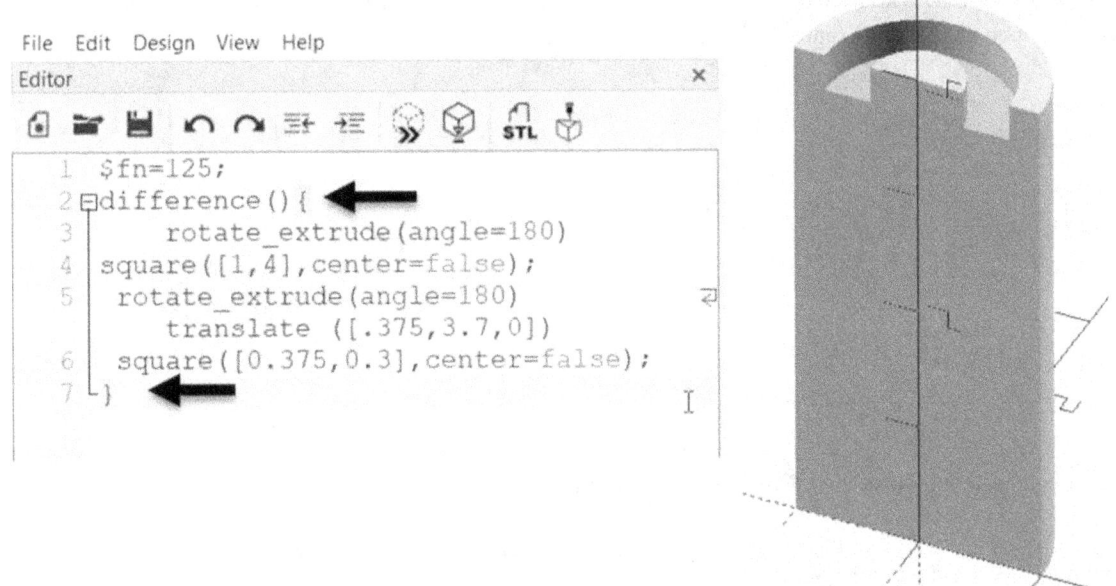

14. Select the entire script except **$fn=125;**. Next, click **Edit > Comment** on the menu bar.

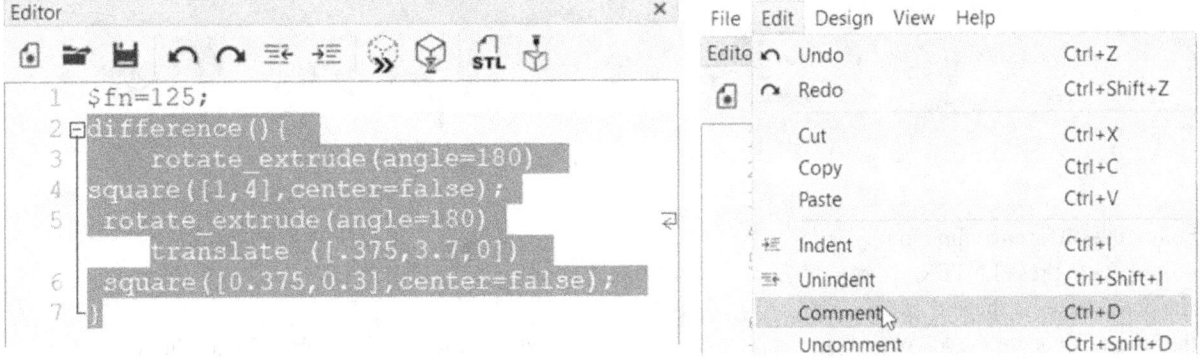

15. Create and translate the rectangle, as shown.

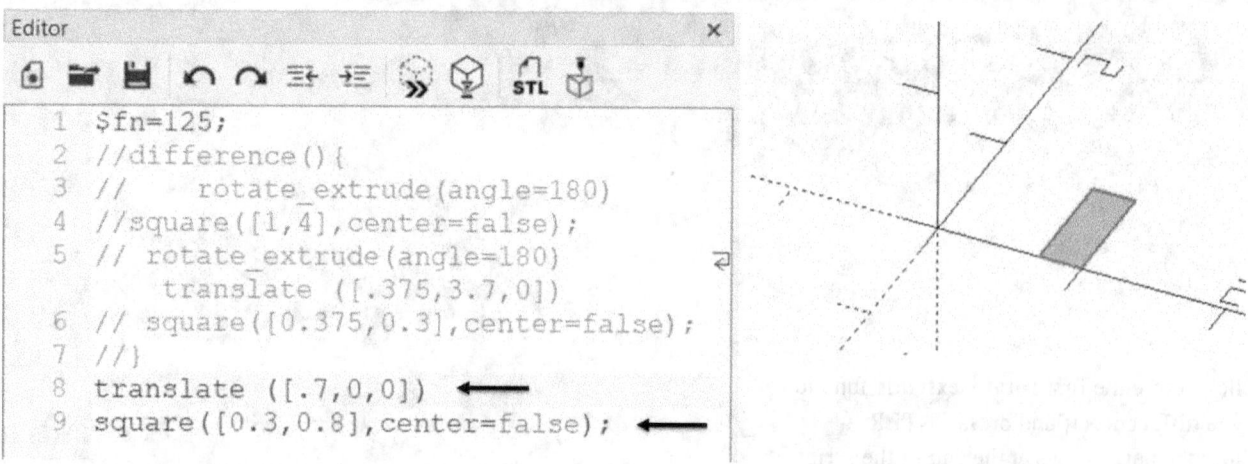

16. Click at the beginning of the **translate** function. Next, type **rotate_extrude(angle=-180)** and press ENTER.
17. Click the **Render** icon on the **View toolbar**; the square is rotate extruded by 180 degrees.

```
1  $fn=125;
2  //difference(){
3  //      rotate_extrude(angle=180)
4  //square([1,4],center=false);
5  //  rotate_extrude(angle=180)
        translate ([.375,3.7,0])
6  // square([0.375,0.3],center=false);
7  //}
8  rotate_extrude(angle=-180)
9  translate ([.7,0,0])
10 square([0.3,0.8],center=false);
```

18. Select the commented script. Next, click **Edit > Uncomment** on the menu bar.

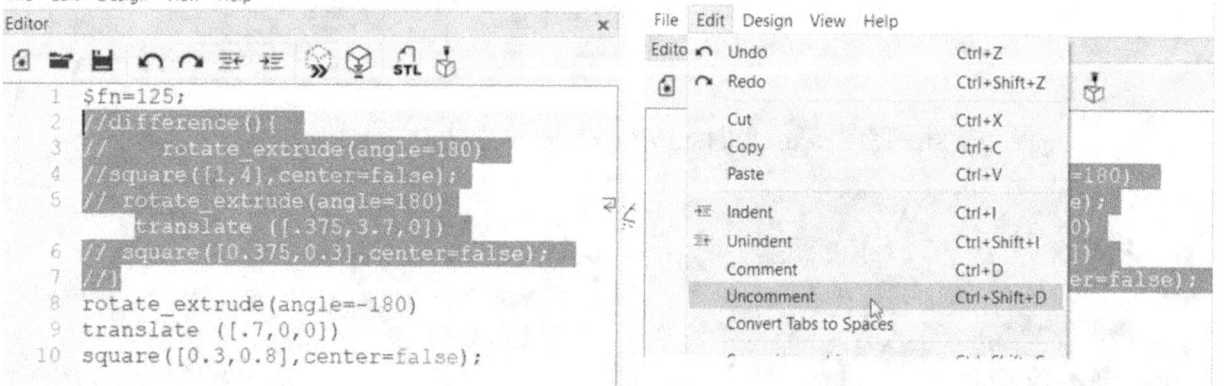

19. Click before the **difference** function.
20. Type **union(){** and press ENTER.
21. Close the parentheses at the end of the script.
22. Click the **Render** icon on the **View** toolbar; the newly created rotate extruded feature is combined with the rest of the model.

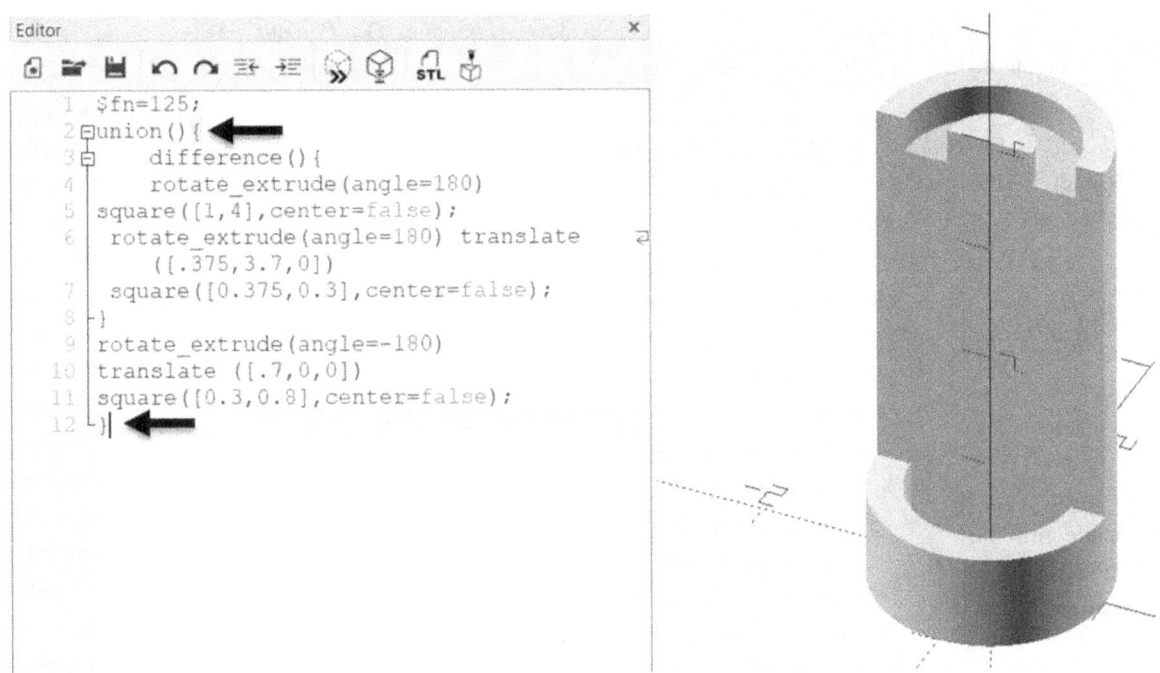

```
Editor                                          ×
1  $fn=125;
2  union(){
3      difference(){
4      rotate_extrude(angle=180)
5  square([1,4],center=false);
6   rotate_extrude(angle=180) translate
        ([.375,3.7,0])
7   square([0.375,0.3],center=false);
8  }
9  rotate_extrude(angle=-180)
10 translate ([.7,0,0])
11 square([0.3,0.8],center=false);
12 }|
```

9. Click the **Save** icon on the **File** toolbar. Next, browse to the required location on your computer.
10. Type **Tutorial_2** in the **File name** box and click **Save**.
11. Click **File > Close** on the menu bar.

Tutorial 3

In this example, you create the part shown below.

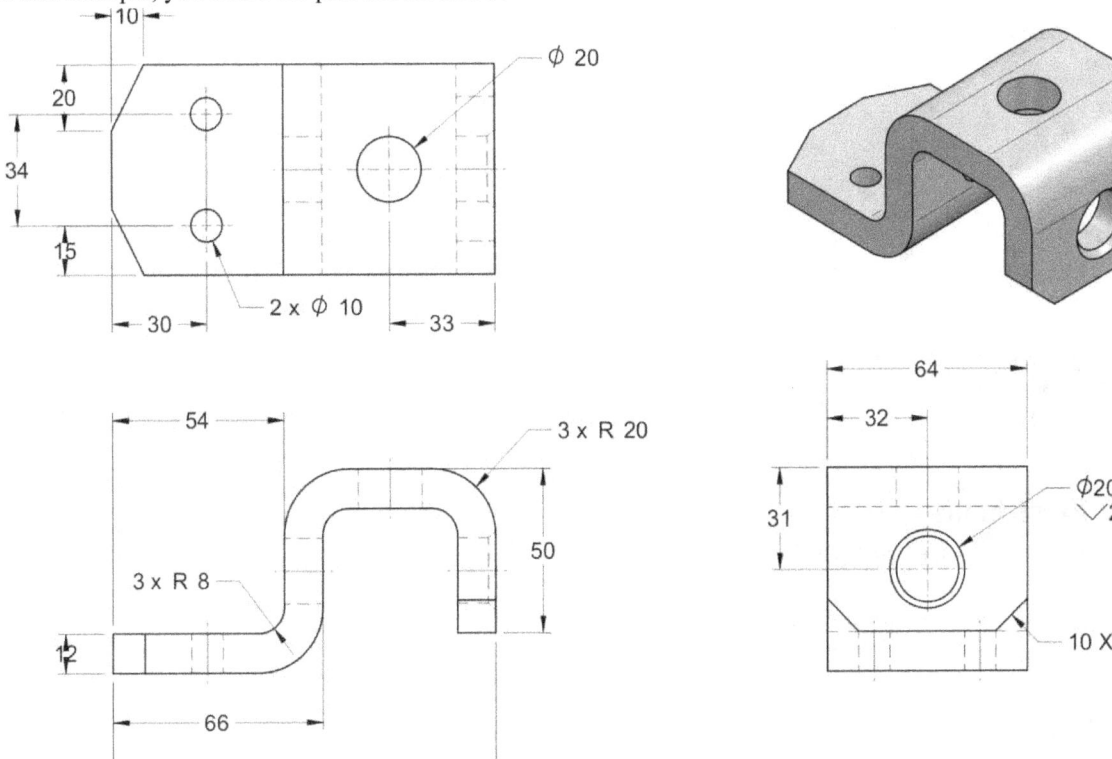

Creating a New document

1. Click the **Windows** icon located at the bottom left corner of the desktop. Next, click **O > OpenSCAD**.
2. Click the **New** button on the **Welcome to OpenSCAD** dialog.

Defining the Resolution and parameters of the model

1. Type **$fn=125;** to define the resolution of the model.
2. Likewise, enter the other parameter values, as shown.

```
$fn=125;
inner_radius = 8;
bend_radius = 20;
bend_angle = 90;
thickness = 12;
height =64;
first_armlength=46;
second_armlength=20;
third_armlength=26;
fourth_armlength=30;
```

Creating the First Arm

1. Type

```
//first arm
cube([first_armlength,height,thickness], center=false);
```

Next, click the **Render** icon to create a cube.

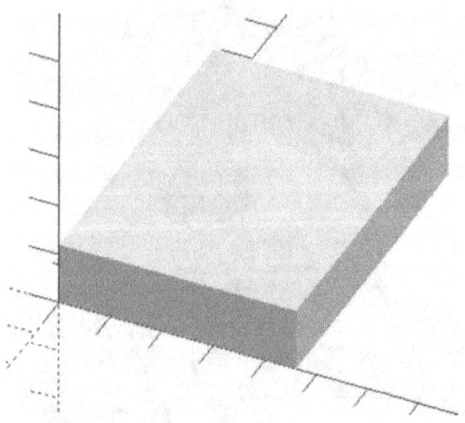

Creating the First bend

Now, you need to create a square and use the **rotate_extrude** function to create the revolved feature.

1. Enter the following functions in the **Editor** window. Next, click the **Render** icon.

```
translate([inner_radius, 0, 0])
square([thickness,height],center=false);
```

A square is created and translated along the X-direction.

2. Click at the beginning of the **translate** function and type **rotate_extrude(angle=bend_angle)**

```
      ralse);
   8  rotate_extrude(angle=bend_angle)  ⬅
   9  translate([inner_radius, 0, 0])
  10  square([thickness,height],center=false);
```

3. Click the **Render** icon; the square is revolved up to the **bend_angle**.

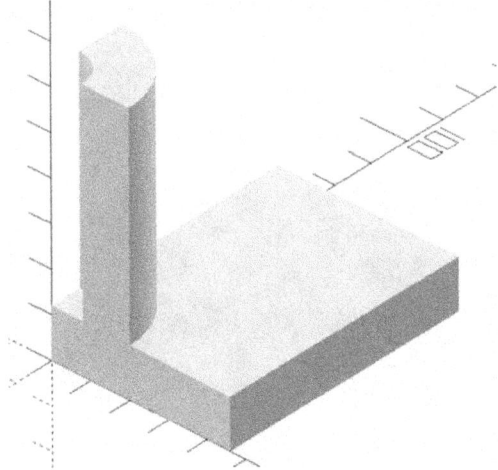

4. Click at the beginning of the **rotate_extrude** function and type **rotate([-90,0,0]){**. Next, press ENTER.
5. Close the loop.

```
   8 ⊟rotate([-90,0,0]){  ⬅
   9       rotate_extrude(angle=bend_angle)
  10  │ translate([inner_radius, 0, 0])
  11  │ square([thickness,height],center=false);  │>
  12  └}  ⬅
```

6. Click the **Render** icon; the revolved feature is rotated by -90 about the X-axis.

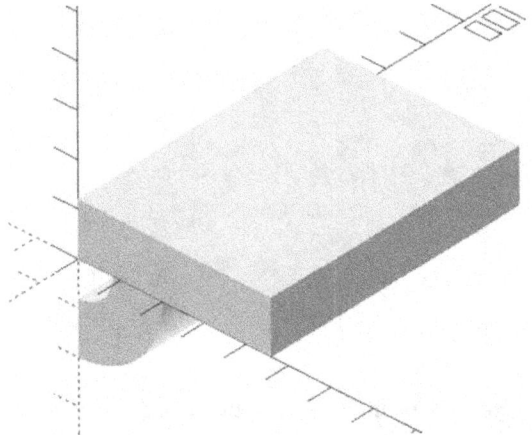

7. Click at the beginning of the **rotate** function and type **translate([first_armlength, 0, bend_radius])**. Next, press ENTER.

```
14  translate([first_armlength, 0, bend_radius])
15 ⊟rotate([-90,0,0]){
16      rotate_extrude(angle=bend_angle)
17 | translate([inner_radius, 0, 0])
18 | square([thickness,height],center=false);
19 └}
```

8. Click the **Render** icon; the revolved feature is translated up to **first_armlength** and **bend_radius** distances in the X and Z-directions, respectively.

9. Click at the beginning of the **translate** function and type **//first_bend**. Next, press ENTER. The bend comment is added to the script below. It will be easy to recognize and modify the values of the bend. You can also reuse the code to create more instances of the bend.

```
11
12  cube([first_armlength,height,thickness],
        center=false);
13  //first_bend
14  translate([first_armlength, 0, bend_radius])
15 ⊟rotate([-90,0,0]){
16      rotate_extrude(angle=bend_angle)
17 | translate([inner_radius, 0, 0])
18 | square([thickness,height],center=false);
19 └}
```

Creating the Second arm

1. Enter the following functions at the end of the script and press ENTER.

    //second_arm
    translate([first_armlength+inner_radius, 0, bend_radius])
    cube([thickness,height,second_armlength], center=false);

2. Click the **Render** icon; a cube is created. Also, it is translated up to 54 distance in the X-direction and distance of the **bend_radius** value in the Z-direction.

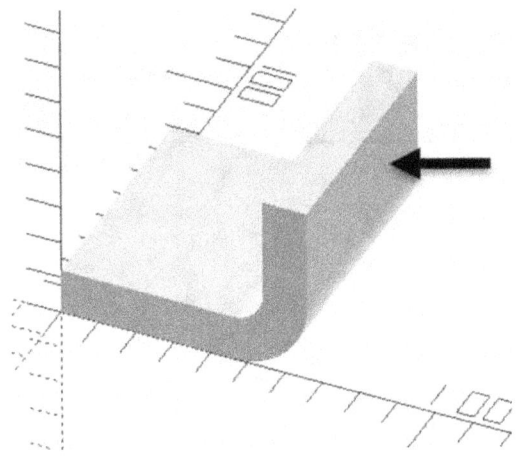

Creating the Second bend

1. Copy the **first_bend** code and paste it at the end of the script.

```
10  fourth_armlength=30;
11
12  cube([first_armlength,height,thickness], center=false);
13  //first_bend
14  translate([first_armlength, 0, bend_radius])
15  rotate([-90,0,0]){
16      rotate_extrude(angle=bend_angle)
17  translate([inner_radius, 0, 0])
18  square([thickness,height],center=false);
19  }
20  translate([first_armlength+inner_radius, 0, bend_radius])
21  cube([thickness,height,27], center=false);
```

← Copy

2. Change the comment to **//second_bend**.

```
20  translate([first_armlength+inner_radius, 0, bel
21  cube([thickness,height,27], center=false);
22  //second_bend  ←
23  translate([first_armlength, 0, bend_radius])
24  rotate([-90,0,0]){
25      rotate_extrude(angle=bend_angle)
26  translate([inner_radius, 0, 0])
27  square([thickness,height],center=false);
28  }
```

3. In the **translate** function, change the distance in the X-direction to **first_armlength +bend_radius+inner_radius**.
4. In the **translate** function, change the distance in the Z-direction to **second_armlength+bend_radius**.
5. In the **rotate** function, change the rotation angle about the Y-axis to **-180**.

```
22  //second_bend
23  translate([first_armlength+bend_radius+inner_radius, 0,
        bend_radius+second_armlength])
24  rotate([-90,-180,0]){
25      rotate_extrude(angle=bend_angle)
26  translate([inner_radius, 0, 0])
27  square([thickness,height],center=false);
28  }
```

6. Click the **Render** icon; the bend is created above the second arm.

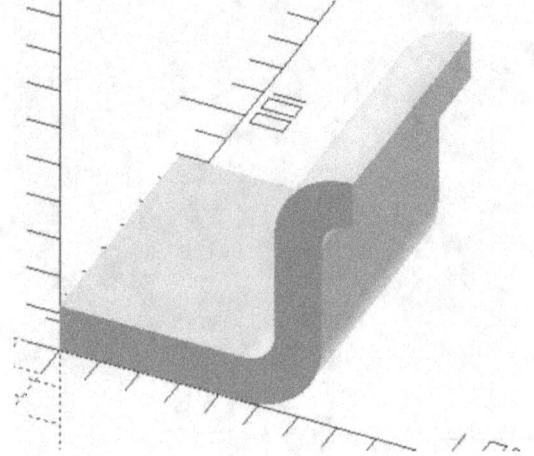

Creating the Third arm

1. Enter the following functions at the end of the script and press ENTER.

//third_arm
translate([first_armlength+inner_radius+bend_radius, 0, bend_radius+second_armlength+inner_radius])
cube([third_armlength,height,thickness], center=false);

```
28  }
29  translate([first_armlength+inner_radius+bend_radius, 0,
        bend_radius+second_armlength+inner_radius])
30  cube([third_armlength,height, thickness], center=false);
31
```

2. Click the **Render** icon; a cube is created.

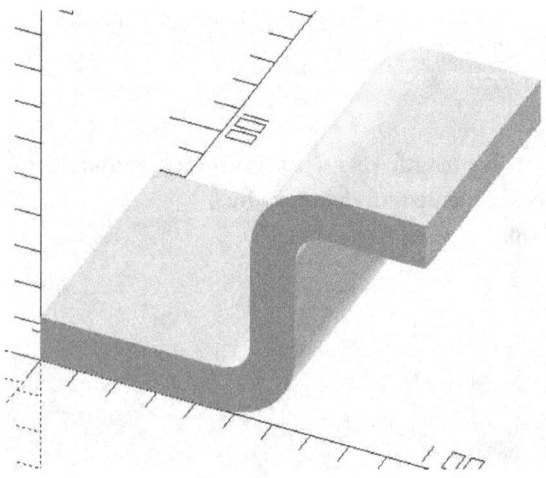

28

Also, the cube is translated in the X-direction up to the distance equal to the sum of **first_armlength**, **bend_radius**, and **inner_radius** of the second bend.

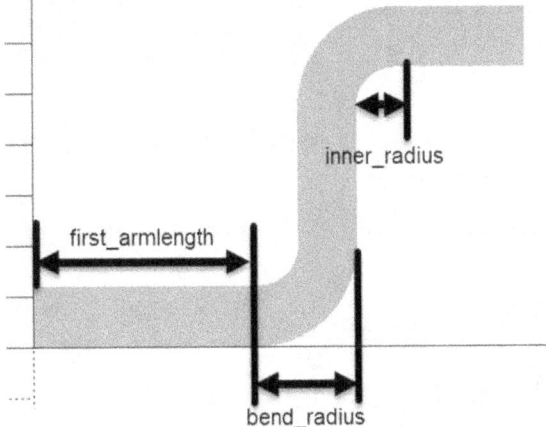

In the Z-direction, the translation distance is the **bend_radius**, **second_armlength**, and **inner_radius** of the second bend.

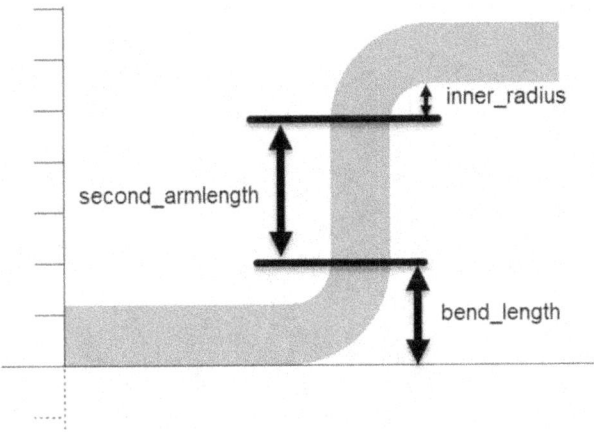

Creating the Third bend

1. Copy the **second_bend** code and paste it at the end of the script.

```
22  //second bend
23  translate([first_armlength+bend_radius+inner_radius, 0,
        bend_radius+second_armlength])
24  rotate([-90,-180,0]){
25      rotate_extrude(angle=bend_angle)            ← Copy
26      translate([inner_radius, 0, 0])
27      square([thickness,height],center=false);
28  }
29  translate([first_armlength+inner_radius+bend_radius, 0,
        bend_radius+second_armlength+inner_radius])
30  cube([third_armlength,height, thickness], center=false);
31
```

2. Change the comment to **//third_bend**.

```
30  cube([third_armlength,height, thickness], center=false);
31  //third_bend ←
32  translate([first_armlength+bend_radius+inner_radius, 0,
       bend_radius+second_armlength])
33  rotate([-90,-180,0]){
34      rotate_extrude(angle=bend_angle)
35  translate([inner_radius, 0, 0])
36  square([thickness,height],center=false);
37  }
```

3. In the **translate** function, change the distance in the X-direction to **first_armlength +bend_radius+inner_radius+third_armlength**.

4. In the **rotate** function, change the rotation angle about the Y-axis to **-90**.

```
31  //third bend
32  translate([first_armlength+bend_radius+
       inner_radius+third_armlength, 0,
       bend_radius+second_armlength])
33  rotate([-90,-90,0]){
34      rotate_extrude(angle=bend_angle)
35  translate([inner_radius, 0, 0])
36  square([thickness,height],center=false);
37  }
38
```

5. Click the **Render** icon; the bend is created next to the third arm.

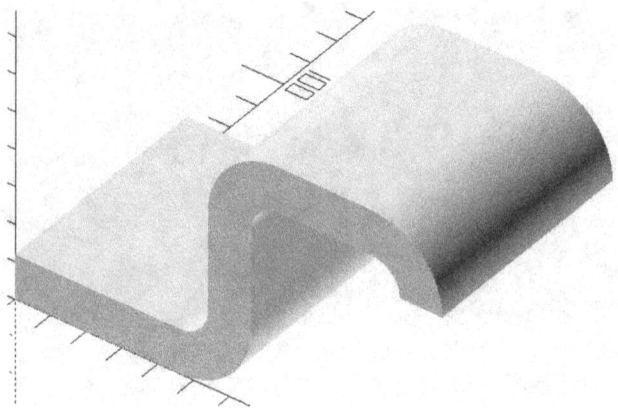

Creating the Third arm

1. Enter the following functions at the end of the script and press ENTER.

   //fourth_arm
   translate([first_armlength+inner_radius+2*bend_radius+third_armlength, 0,
   bend_radius+second_armlength])
   rotate([0,180,0])
   cube([thickness,height,fourth_armlength], center=false);

3. Click the **Render** icon; a cube is created.

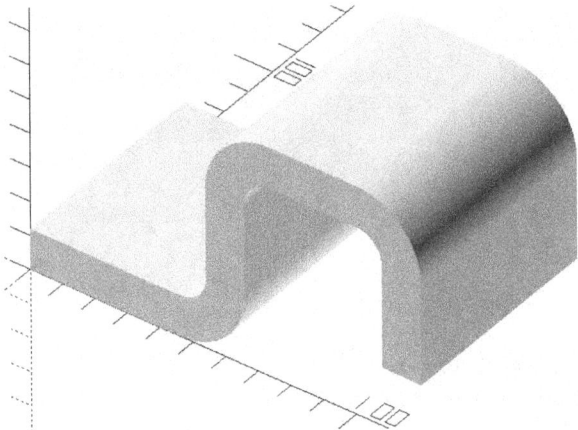

Also, the cube is translated in the X-direction up to the distance equal to the sum of **first_armlength**, **2 X bend_radius**, **inner_radius** of the second bend, and **third _armlength**.

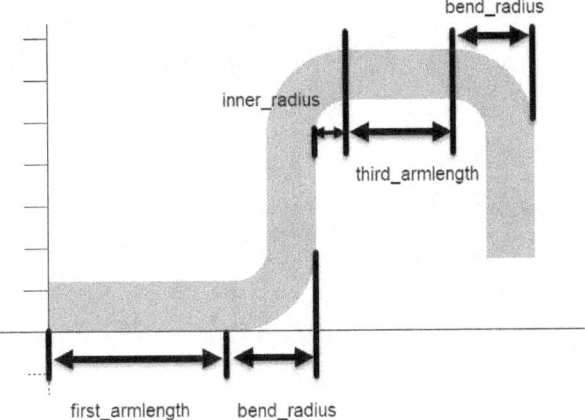

In the Z-direction, the translation distance is the **bend_radius**, and **second_armlength**.

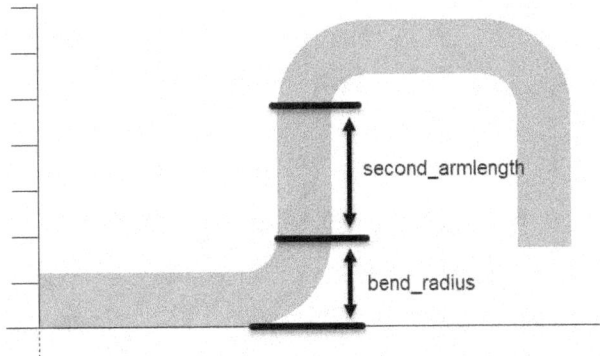

4. Add the **union(){** Boolean operation before the **//first arm** comment. Next, press ENTER.

```
 9  third_armlength=26;
10  fourth_armlength=30;
11  union(){
12  //first_arm
13  cube([first_armlength,height,
        thickness], center=false);
14  //first_bend
```

5. Close the loop at the end of the script, as shown.

```
43  rotate([0,180,0])
44  cube([thickness,height,
        fourth_armlength], center=f
45  }
```

Creating the Holes

1. Type **//chamfered_hole** at the end of the script. Next, press ENTER.
2. Type **cylinder(66,d=21,false);** to create a cylinder of 66 height and 21 diameter.
3. Press ENTER and type **cylinder(h=1.3, r1=10.5, r2=12, center=false);**
4. Click the **Render** icon; another cylinder with a height of 1.3, a start radius of 10.5, and an end radius of 12 is created.

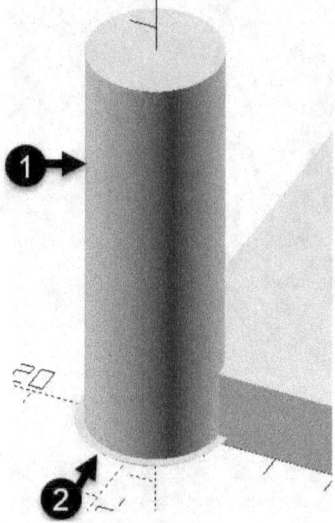

5. Click before the second cylinder and type **translate([0,0,64.7])**. The second cylinder is translated up to a distance of 64.7 in the Z-direction.

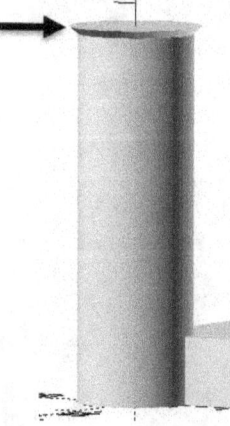

6. Click before the first cylinder and **rotate([0,90,0]){**.
7. Next, close the parentheses } at the end of the script. The two cylinders are rotated by 90 degrees about the Y-axis.

32

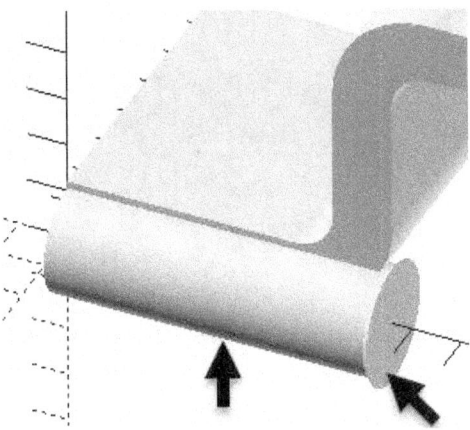

8. Click before the **rotate([0,90,0])** function and type **translate([54,32,33])**. The rotated cylinders are translated.

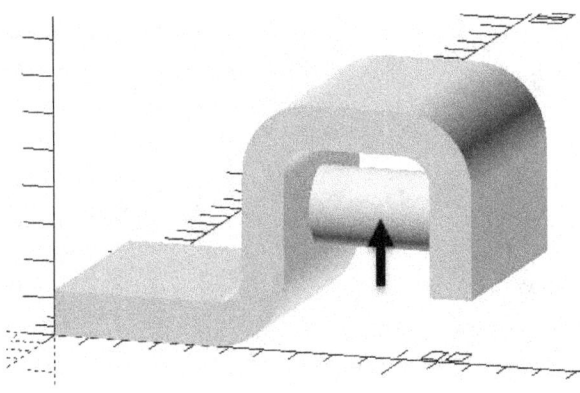

9. Add the **difference(){** Boolean operation before the **union(){** operation. Next, press ENTER.

```
 9  third_armlength=26;
10  fourth_armlength=30;
11  difference(){   ⬅
12  union(){
13  //first_arm
14  cube([first_armlength,height,
         thickness], center=false);
15  //first_bend
16  translate([first_armlength, 0,
         bend_radius])
17  rotate([ 90 0 0]){
```

10. Close the loop at the end of the script, as shown.

```
47  //chamfered_hole
48  translate([54,32,33])
49  rotate([0,90,0]){
50  cylinder( 66,    d=21,false);
51  translate([0,0,64.7])
52  cylinder(h=1.3,  r1=10.5, r2=12,          ⏎
         center=false);
53  }
54  }   ⬅
```

11. Click the **Render** icon to create the chamfered hole.

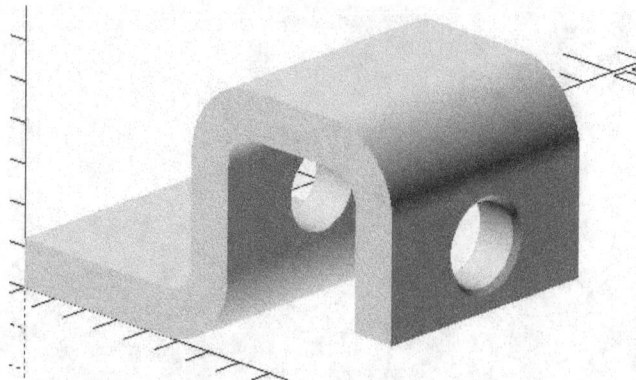

12. At the bottom of the script, click inside the closed-loop and press ENTER.

```
51  translate([0,0,64.7])
52  cylinder(h=1.3, r1=10.5, r2
        center=false);
53  }|
54  }
55
```

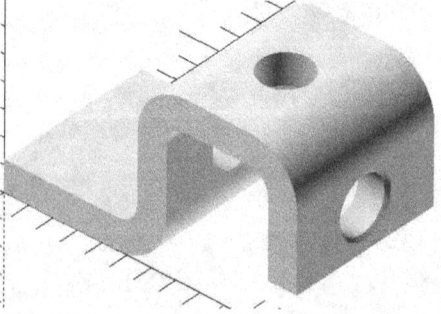

Click here

13. Enter the **//second_hole** comment.
14. Enter the **translate()** and **cylinder ()** functions with the values, as shown.

```
53  }
54  //second_hole
55  translate([87,32,0])
56  cylinder( 74,d=20,false);|
57  }
```

15. Click the **Render** icon to create another hole.

16. Click at the end of the **cylinder(74,d=20,false);** function and press ENTER.
17. Enter the **//third_hole** comment.
18. Enter the **translate()** and **cylinder ()** functions with the values, as shown.

```
57  //third_hole
58  translate([30,15|,0])
59  cylinder( 12,d=10,false);
60  }
61
```

19. Click the **Render** icon to create another hole.

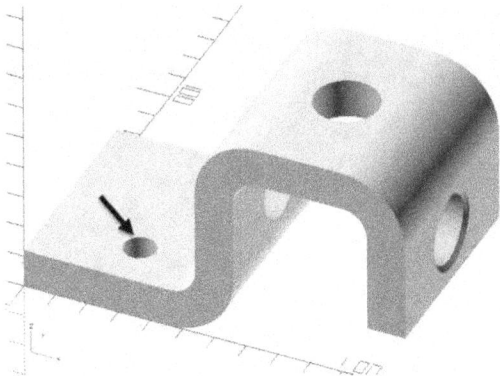

20. Likewise, create the fourth hole by using the cylinder and translate functions.

```
60  //fourth_hole
61  translate([30,45,0])
62  cylinder( 12,d=10,false);
63  }
64
```

Creating Chamfers

1. Click at the end of the **cylinder(12,d=10,false);** function and press ENTER.
2. Enter the **//first_chamfer** comment.
3. Enter the **polygon** function and enter the three points of the polygon in the brackets, as shown.

```
62  cylinder( 12,d=10,false);
63  //first_chamfer
64  polygon([[0,0],[10,0],[0,20]]);  ⬅
65  }
```

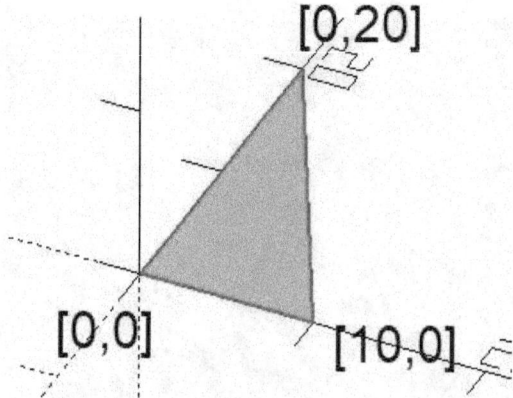

4. Click at the beginning of the **polygon ([[0,0],[10,0],[0,20]]);** function.
5. Type **linear_extrude(height = 12, center = false)** and press ENTER. The polygon is extruded in the linear direction up to 12 distance. Also, the extrusion is not centered.
6. Click the **Render** icon to perform the difference operation between the linear extrusion and the remaining volume.

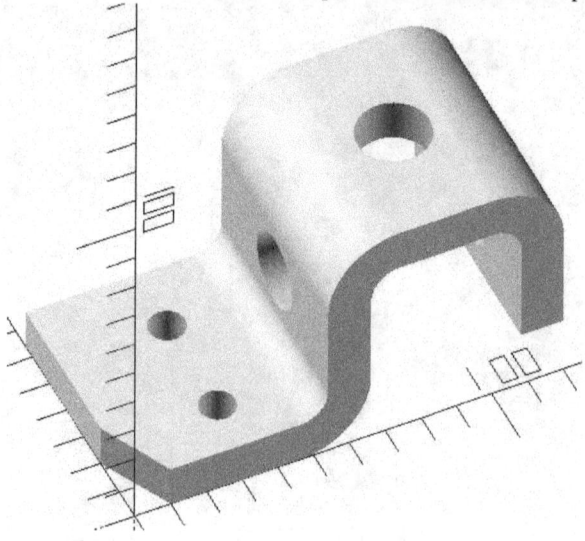

7. Click at the end of the **polygon ([[0,0],[10,0],[0,20]]);** function and press ENTER.
8. Enter the **//second_chamfer** comment.
9. Enter the **linear_extrude** and **polygon** functions, as shown.

```
66 //second_chamfer
67 linear_extrude(height = 12, center =  ⊇
      false)
68 polygon([[0,0],[10,20],[0,20]]);
69 }
```

10. Click at the beginning of the **linear_extrude(height = 12, center = false)** function.
11. Type **translate([0,24,0])** and press ENTER.

```
66 //second_chamfer
67 translate([0,44,0])
68 linear_extrude(height = 12, center =
      false)
69 polygon([[0,0],[10,20],[0,20]]);
70 }
71
```

12. Click the **Render** icon to perform the difference operation between the linear extrusion and the remaining volume.

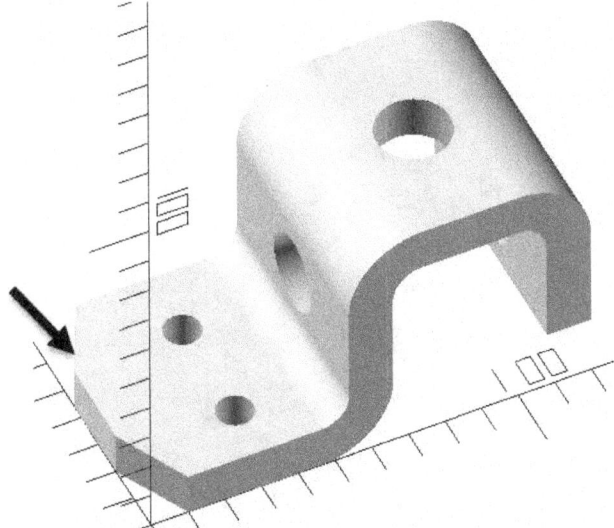

13. Likewise, create two chamfers, as shown.

```
//third_chamfer
translate([120,0,10])
rotate([0,-90,0]){
linear_extrude(height = 12, center = false)
polygon([[0,0],[10,0],[0,10]]);
}
//fourth_chamfer
translate([108,64,10])
rotate([180,-90,0]){
linear_extrude(height = 16, center = false)
polygon([[0,0],[10,0],[0,10]]);
}
```

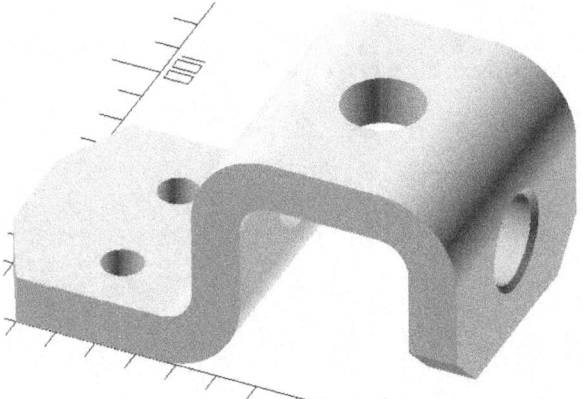

14. Click the **Save** icon on the **File** toolbar. Next, browse to the required location on your computer.
15. Type **Tutorial_1** in the **File name** box and click **Save**.
16. Click **File > Close** on the menu bar.

Tutorial 4

In this example, you create the part shown next.

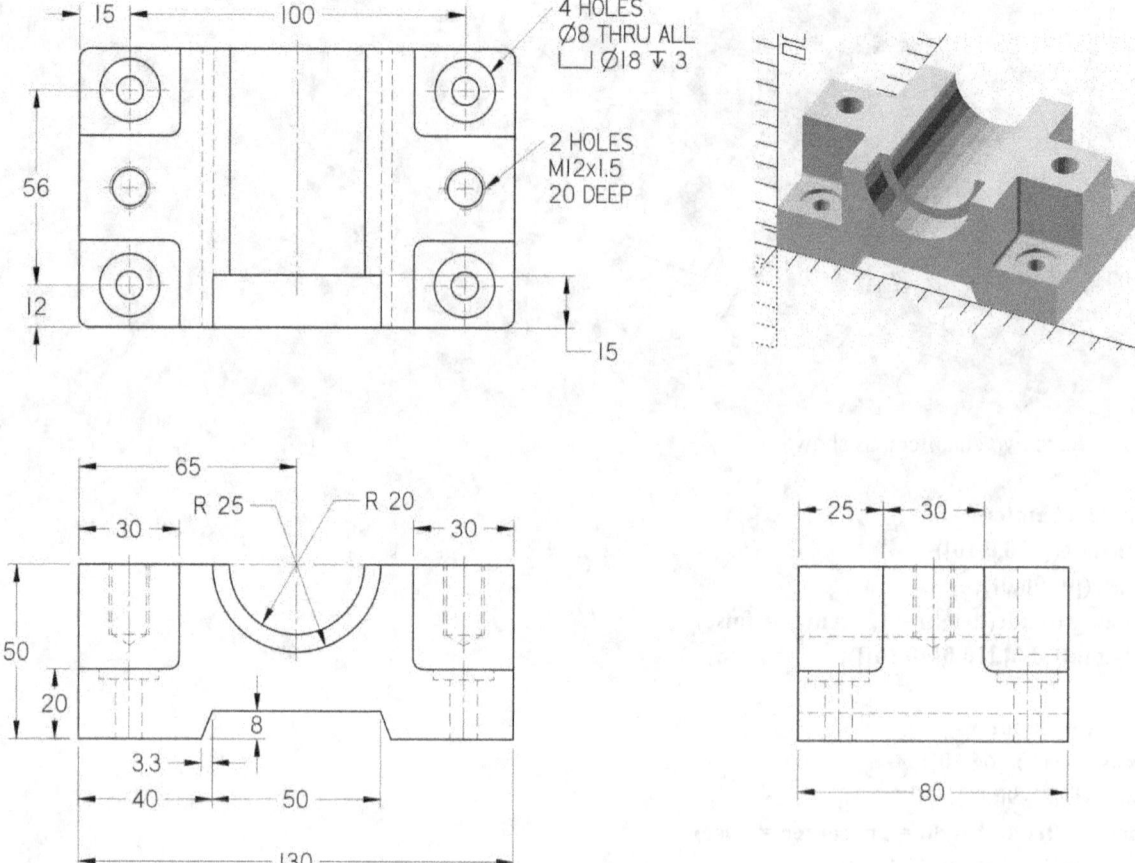

Creating a New document

1. Click the **Windows** icon located at the bottom left corner of the desktop. Next, click **O > OpenSCAD**.
2. Click the **New** button on the **Welcome to OpenSCAD** dialog.

Creating the Body

1. Type **$fn=125;** to define the resolution of the model.
2. Type **cube([130,80,50]);** and click the **Render** icon.

```
1  $fn=25;
2  cube([130,80,50]);
```

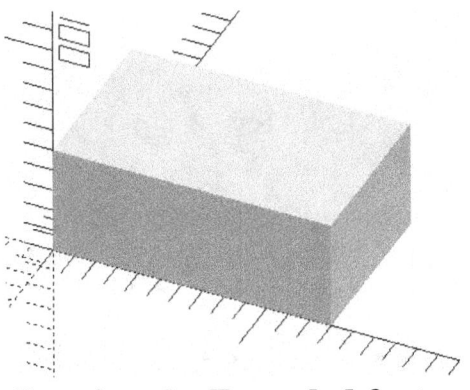

Creating the Extruded features

1. Click the **New** icon on the **File toolbar**.

2. Type **cube([30,25,30]);** and click the **Render** icon.

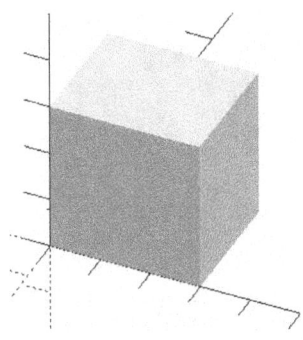

Next, you need to add fillets to the edges of the cube. You can add fillets using the **sphere ()** and **minkowski()** **functions.**

3. Type **sphere(d=4);** below the **cube()** function.
4. Click the **Render** icon. A sphere of 4 diameter is created.

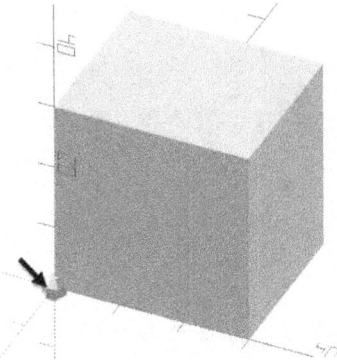

Next, you need to apply the **minkowski()**function between the cube and sphere.

5. Click at the beginning of the **cube()** function.
6. Type **minkowski(){**.
7. Next add **}** at the end of the **sphere()** function.

```
1  minkowski(){cube([30,25,30]);
2  sphere(d=4);
3  }
```

8. Click the **Render** icon. The sphere is transformation along all the edges of the cube. As a result, the edges are rounded. Also, notice that the size of the cube is increased by the radius of the sphere on all sides.

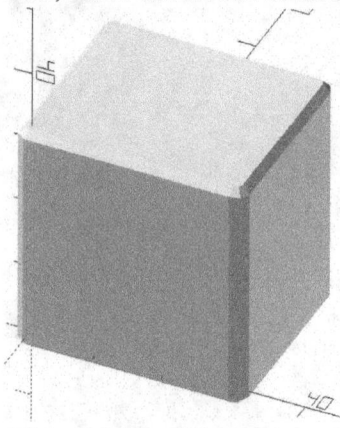

9. Click at the beginning of the **minkowski()** function.
10. Type **translate([-2,-2,22])** and press ENTER.
11. Click the **Render** icon.

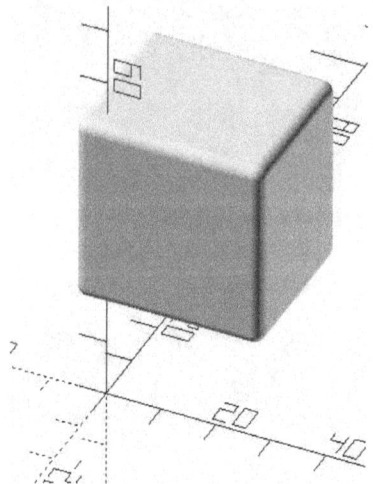

Next, you need to remove the material from the main body using the filleted cube.

12. Select the entire script. Next, right-click and select **Copy**.

13. Switch to the window of the main part.
14. Right-click below the **cube()** function, and then select **Paste**.

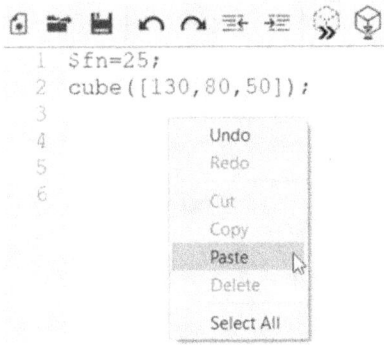

15. Click at the beginning of the **cube()** function.
16. Type **difference(){** and press ENTER.
17. Next, add **}** at the end of the script.

```
1  $fn=25;
2  difference(){
3       cube([130,80,50]);
4  translate([-2,-2,22])
5  minkowski(){cube([30,25,30]);
6  sphere(d=4);
7  }
8  }
9
```

18. Click the **Render** icon. The filleted cube is removed from the main body.

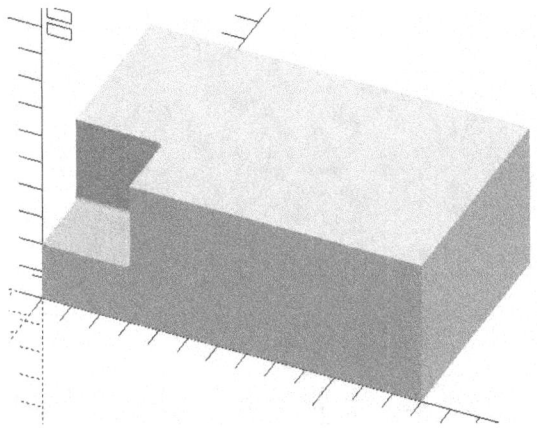

Creating the Rectangular Pattern of the Cut Features

In OpenSCAD, you can pattern the objects using the **For** loop and **translate** function.

41

1. Click at the end of the **cube([130,80,50]);** function, and then press ENTER.

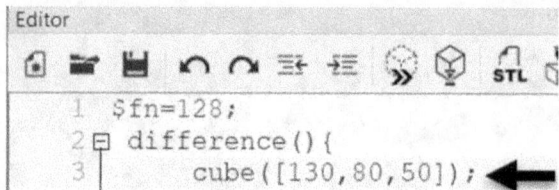

2. Type **for (x=[0:104:104]){** and press ENTER. This defines the **x** variable and its parameters.

   x = variable
   0 = start (initial position)
   104 = increment
   104 = end (end position)

3. Type **translate([x,0,0]){** and press ENTER.
4. Close the **for** and **translate** loops.

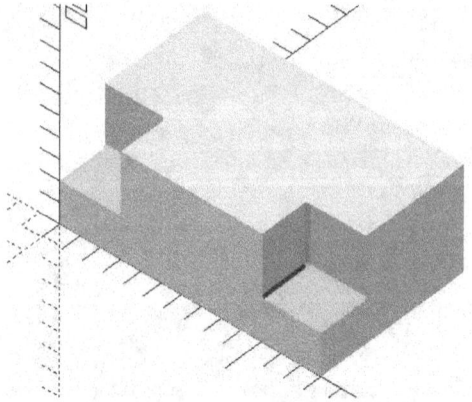

5. Click the **Render** icon. The rounded cut feature is patterned along the x axis.

Next, you need to pattern the cut feature in the Y-direction.

6. Click after **for (x=[0:104:104]){** and press ENTER.

```
Editor
1   $fn=5;
2 ⊟ difference(){
3        cube([130,80,50]);
4 ⊟ for (x=[0:104:104]){  ◄ Click Here
5 ⊟          translate([x,0,0]){
       translate([-2,-2,22])
6 ⊟minkowski(){cube([30,25,30]);
7  sphere(d=4);
8 ⊢}
9 ⊢}
```

7. Type **for (y=[0:58:58]){** and press ENTER. It defines the **y** variable and its parameters.

    y = variable
    0 = start (initial position)
    58 = increment
    58= end (end position)

8. Change the **Y** coordinate value of the **translate** function to **y**.
9. Close the new **for** loop.

```
Editor
1   $fn=5;
2 ⊟ difference(){
3        cube([130,80,50]);
4 ⊟ for (x=[0:104:104]){
5 ⊟    for (y=[0:58:58]){
6 ⊟           translate([x,y,0]){
       translate([-2,-2,22])
7 ⊟minkowski(){cube([30,25,30]);
8  sphere(d=4);
9 ⊢}
10 ⊢}
11 ⊢}
12 ⊢}
13  }
14
```

10. Click the **Render** icon. The rounded cut feature is patterned along the y axis.

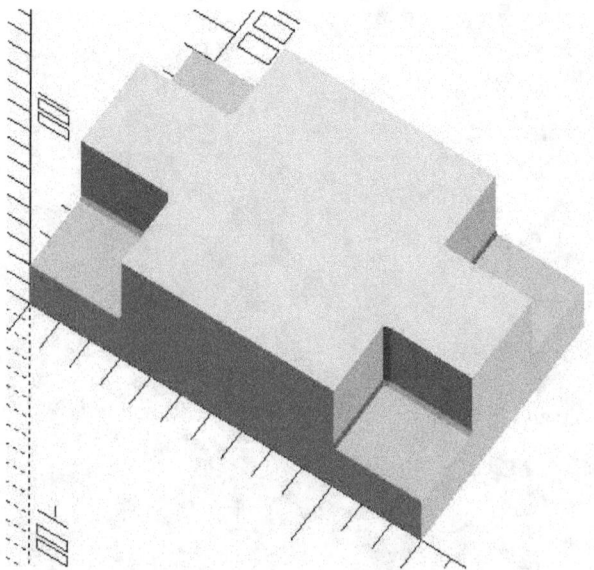

Creating the Hole features

1. Click the **New** icon on the **File** toolbar.

2. Enter the following script to created two cylinders.

    ```
    cylinder(18,d=8,false);
    translate([0,0,18])
    cylinder(3,d=18,false);
    ```

3. Click the **Render** icon.

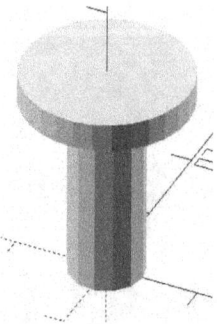

4. Click at the beginning of the **cylinder()** function.
5. Type **translate([15,12,0]){** and press ENTER.
6. Next, add **}** at the end of the script.

    ```
    translate([15,12,0]){
       cylinder(18,d=8,false);
    translate([0,0,18])
    ```

44

```
cylinder(3,d=18,false);
}
```

7. Click the **Render** icon. The two cylinders are translated in the X and Y directions.

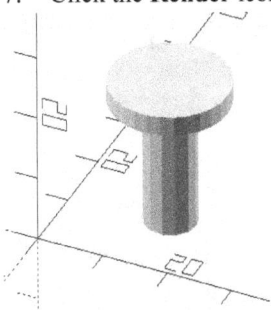

8. Add the **for** and **translate** loops to create the rectangular pattern of the cylinders.

```
for (x=[0:100:100]){
    for (y=[0:56:56]){
        translate([x,y,0]){
```

9. Close the **for** and **translate** loops.

```
1 for (x=[0:100:100]){
2     for (y=[0:56:56]){
3         translate([x,y,0]){
        translate([15,12,0]){
        cylinder(18,d=8,false);
4 translate([0,0,18])
5 cylinder(3,d=18,false);
6 }
7 }
8 }
9 }
```

10. Click the **Render** icon.

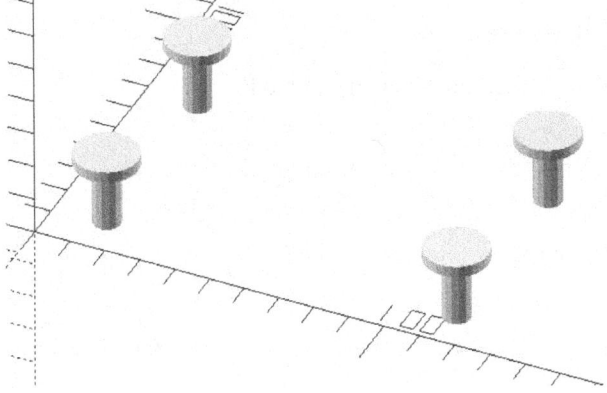

11. Select the entire script. Next, right-click and select **Copy**.
12. Switch to the window of the main part.

13. Click at the closing parentheses of the **for** loop, and then press ENTER.

```
1  $fn=5;
2  difference(){
3      cube([130,80,50]);
4  for (x=[0:104:104]){
5      for (y=[0:58:58]){
6              translate([x,y,0]){
       translate([-2,-2,22])
7  minkowski(){cube([30,25,30]);
8  sphere(d=4);
9  }
10 }
11 }
12 } I   ⬅
13 }
14
```

14. Right-click, and then select **Paste**.

```
1  $fn=5;
2  difference(){
3      cube([130,80,50]);
4  for (x=[0:104:104]){
5      for (y=[0:58:58]){
6              translate([x,y,0]){
       translate([-2,-2,22])
7  minkowski(){cube([30,25,30]);
8  sphere(d=4);
9  }
10 }
11 }
12 }
3  for (x=[0:100:100]){
4      for (y=[0:56:56]){
5          translate([x,y,0]){translate([15,
           12,0]){cylinder(18,d=8,false);
6  translate([0,0,18])
7  cylinder(3,d=18,false);
8  }
9  }
0  }
1  }
22 }
```

15. Click the **Render** icon. The cylinders are removed from the main body resulting in counterbored holes.

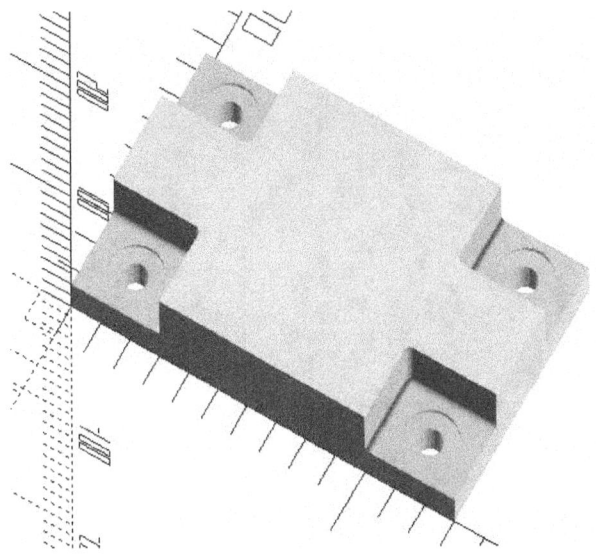

16. Click the **New** icon on the **File** toolbar.
17. Enter the following script to created two cylinders.

```
cylinder(65,d=40,false);
translate([0,0,65])
cylinder(15,d=50,false);
```

18. Rotate the cylinders using the **rotate** function.

```
rotate([90,0,0]){
cylinder(65,d=40,false);
translate([0,0,65])
cylinder(15,d=50,false);
}
```

19. Translate the cylinders by a distance of 65, 80, and 50 along the X, Y, and Z directions, respectively.

```
translate([65,80,50])
rotate([90,0,0]){
cylinder(65,d=40,false);
translate([0,0,65])
cylinder(15,d=50,false);
}
```

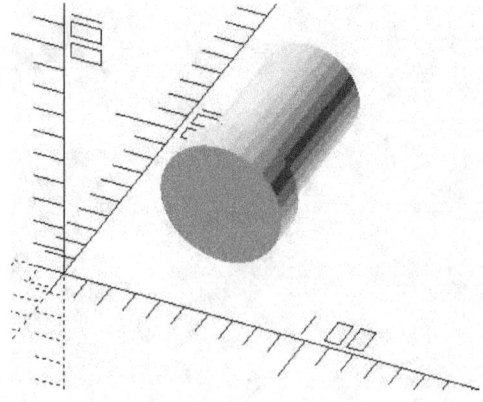

20. Select the entire script. Next, right-click and select **Copy**.
21. Switch to the window of the main part.
22. Click after the closing parentheses of the **for** loop, and then press ENTER.

```
12 }
13 for (x=[0:100:100]){
14     for (y=[0:56:56]){
15         translate([x,y,0]){    ⇄
           translate([15,12,0]){  ⇄
           cylinder(18,d=8,false  ⇄
           );
16 translate([0,0,18])
17 cylinder(3,d=18,false);
18 }
19 }
20 }
21 } ⬅
22 }
```

23. Right-click, and then select **Paste**.

```
12 }
13 for (x=[0:100:100]){
14     for (y=[0:56:56]){
15         translate([x,y,0    ⇄
           ]){translate([15,12 ⇄
           ,0]){cylinder(18,d=  ⇄
           8,false);
16 translate([0,0,18])
17 cylinder(3,d=18,false);
18 }
19 }
20 }
21 }
22 translate([65,80,50])
23 rotate([90,0,0]){
24 cylinder(65,d=40,false);
25 translate([0,0,65])
26 cylinder(15,d=50,false);
27 }
28 }
```

24. Click the **Render** icon on the **File toolbar**.

48

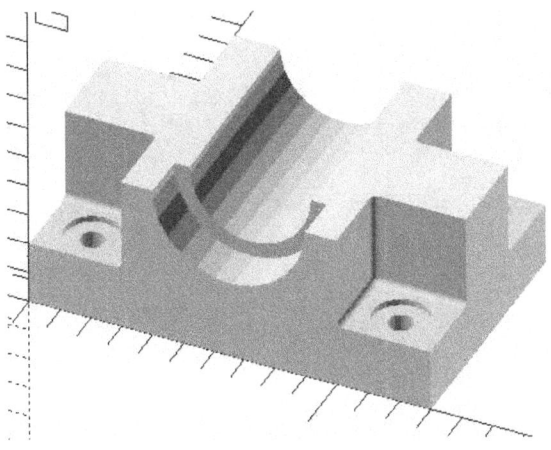

25. Click the **New** icon on the **File** toolbar.
26. Type **cylinder(h=3, r1=0, r2=6, center=false);** and press ENTER.

    h=height of the cylinder
    r1=start radius of the cylinder
    r2=end radius of the cylinder.

    Next, you need to create another cylinder and translate it up to 3 distance in the Z direction.

27. Type **translate([0,0,3])** and press ENTER.
28. Type **cylinder(20,d=12,false);**

Height of the cylinder = 20
Diameter = 12.

29. Click the **Render** icon.

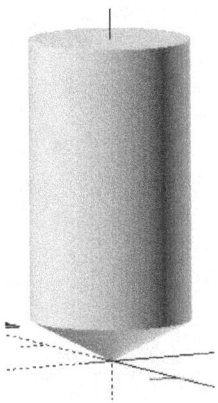

30. Translate the cylinders by a distance of 15, 40, and 30 along the X, Y, and Z directions, respectively.

```
translate([15,40,30]){
cylinder(h=3, r1=0, r2=6, center=false);
translate([0,0,3])
cylinder(20,d=12,false);
```

}

31. Click the **Render** icon.

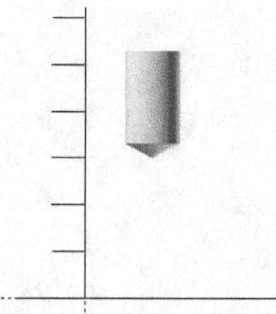

32. Add the **for** and **translate** loops to create the linear pattern of the cylinders along the X-direction.

    for (x=[0:100:100]){
      translate([x,0,0]){

33. Close the **for** and **translate** loops.

```
2  for (x=[0:100:100]){
3      translate([x,0,0]){
4  translate([15,40,30]){
5  cylinder(h=3, r1=0, r2=6,
          center=false);
6  translate([0,0,3])
7  cylinder(20,d=12,false);
8  }
9      }
10     }
```

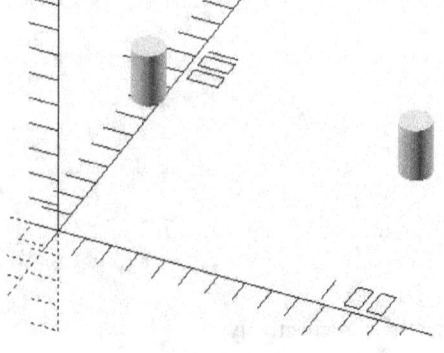

34. Click the **Render** icon.

35. Select the entire script. Next, right-click and select **Copy**.
36. Switch to the window of the main part.
37. Click after the closing parentheses of the **rotate** loop, and then press ENTER.

```
22  translate([65,80,50])
23  rotate([90,0,0]){
24  cylinder(65,d=40,false);
25  translate([0,0,65])
26  cylinder(15,d=50,false);
27  }
28  }
29
```

38. Right-click, and then select **Paste**.

```
22  translate([65,80,50])
23  rotate([90,0,0]){
24  cylinder(65,d=40,false);
25  translate([0,0,65])
26  cylinder(15,d=50,false);
27  }
28  for (x=[0:100:100]){
29      translate([x,0,0]){
30  translate([15,40,30]){
31  cylinder(h=3,  r1=0,  r2=
        6, center=false);
32  translate([0,0,3])
33  cylinder(20,d=12,false);
34  }
35      }
36      }
37  }
```

39. Click the **Render** icon. The cylinders are removed from the main body resulting in a linear pattern of holes.

Creating the Extruded cut feature

1. Click the **New** icon on the **File** toolbar.
2. Enter the **polygon** function and enter the three points of the polygon in the brackets, as shown.

polygon([[36.7,0],[93.3,0],[90,8],[40,8],]);

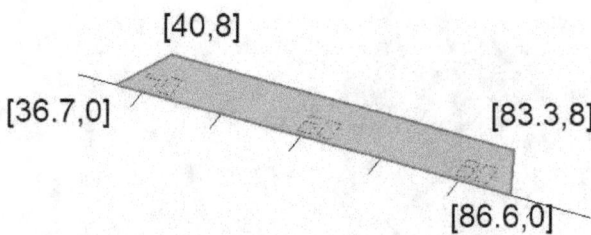

3. Click at the beginning of the **polygon([[36.7,0],[93.3,0],[90,8],[40,8],]);** function.
4. Type **linear_extrude(height = 80, center = false)** and press ENTER. The polygon is extruded in the linear direction up to 80 distance. Also, the extrusion is not centered.
5. Click the **Render** icon.

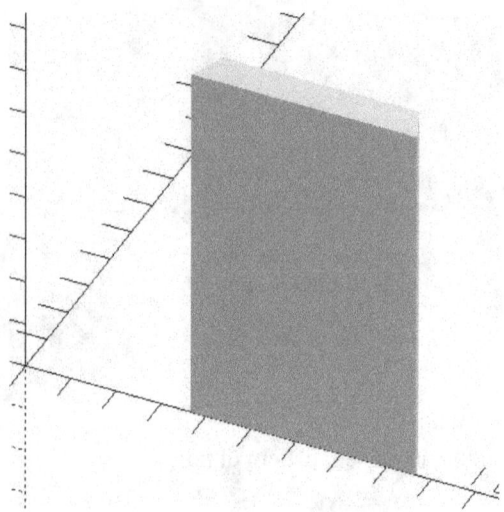

40. Rotate the model using the **rotate** function.

```
rotate([90,0,0])
linear_extrude(height = 80, center = false)
polygon([[36.7,0],[93.3,0],[90,8],[40,8],]);
```

41. Translate the model by a distance of 80 along the Y direction, respectively.

```
translate([0,80,0])
rotate([90,0,0])
linear_extrude(height = 80, center = false)
polygon([[36.7,0],[93.3,0],[90,8],[40,8],]);
```

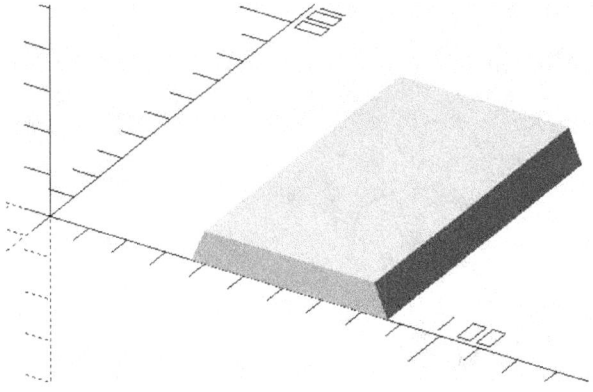

42. Select the entire script. Next, right-click and select **Copy**.
43. Switch to the window of the main part.
44. Click after the closing parentheses of the **for** loop, and then press ENTER.

```
27 }
28 for (x=[0:100:100]){
29     translate([x,0,0]){
30 translate([15,40,30]){
31 cylinder(h=3,  r1=0,  r2=
        6, center=false);
32 translate([0,0,3])
33 cylinder(20,d=12,false);
34 }
35     }
36     }  ⬅
37 }
38
```

45. Right-click, and then select **Paste**.
46. Click the **Render** icon on the **File** toolbar.

47. Click the **Save** icon on the **File** toolbar.
48. Type Tutorial_4 in the **File name** box and click **Save**.
49. Close the file.

Tutorial 5

In this example, you create the part shown below.

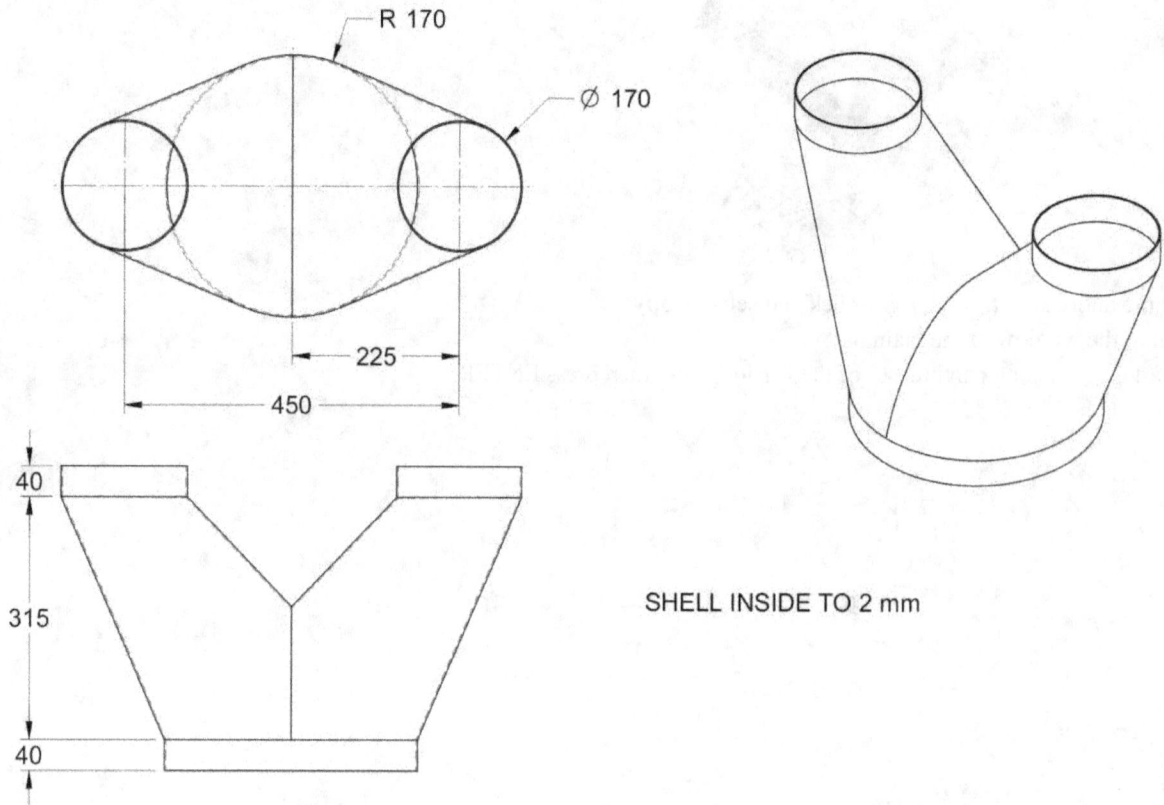

SHELL INSIDE TO 2 mm

Creating a New document

1. Click the **Windows** icon located at the bottom left corner of the desktop. Next, click **O > OpenSCAD**.
2. Click the **New** button on the **Welcome to OpenSCAD** dialog.

Using the Hull function

1. Type the following two parameters.

    v1=340;
    v2=170;
    v3=338;
    v4=168;

2. Type the following script to create two extruded cylinders, as shown.

    translate([0,0,40])
    cylinder(h=0.0001,d=340);
    translate([225,0,355])
    cylinder(h=0.0001,d=170);

3. Click the **Render** icon.

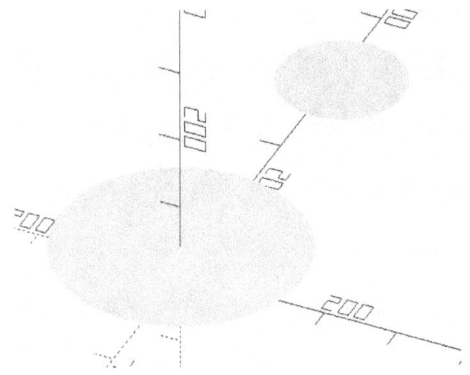

4. Click at the beginning of the first **translate ()** function.
5. Type **hull(){**.
6. Close the loop at the end of the script.

```
hull(){
    translate([0,0,40])
cylinder(h=0.0001,d=d1);
translate([225,0,355])
cylinder(h=0.0001,d=d2);
}
```

7. Click the **Render** icon.

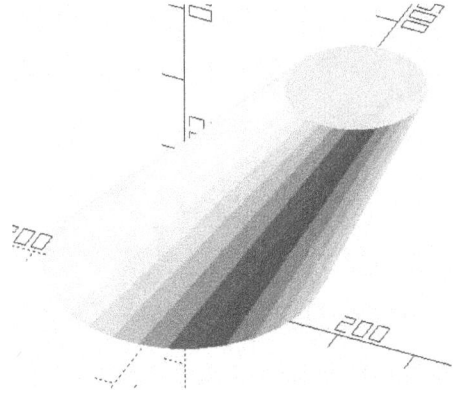

Using a Module

1. Click at the beginning of the **hull(){**function.
2. Press Enter.
3. Type **module loft (d1,d2){**

loft = name of the module
d1 and **d2** are parameters of the module.

4. Change the **d** values of the first and second cylinders to **d1** and **d2**, respectively.
5. Close the loop at the end of the script.

```
module loft (d1,d2){
    hull(){
    translate([0,0,40])
```

```
cylinder(h=0.0001,d= );
translate([225,0,355])
cylinder(h=0.0001, d= );
}
}
```

Now, the module is defined, and you need to use the module to create features.

6. Type **loft (d1=v1,d2=v2);**
7. Click the **Render** icon to create the loft feature.

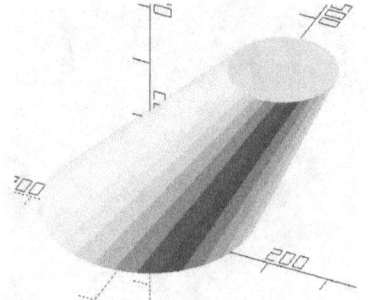

Mirroring the Loft module

1. Type **mirror([1,0,0]) loft (d1=v1,d2=v2);**
2. Click the **Render** icon to mirror the loft feature about the X axis.

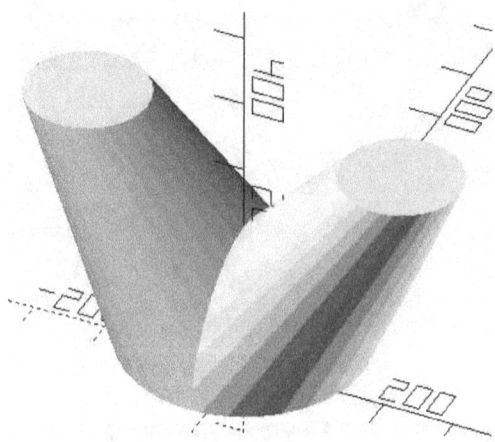

3. Create three cylinders by entering the script, as shown.

```
cylinder(h=40,d=v1);
translate([225,0,355])
cylinder(h=40,d=v2);
translate([-225,0,355])
cylinder(h=40,d=v2);
```

Next, you need to combine all the solids using the **union()**Boolean operation.

4. Click at the beginning of **loft (d1=v1,d2=v2);**
5. Type **union(){** and press ENTER.
6. Close the loop at the end of the script.

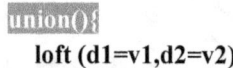
```
    loft (d1=v1,d2=v2);
```

56

```
mirror([1,0,0]) loft (d1=v1,d2=v2);
cylinder(h=40,d=v1);
translate([225,0,355])
cylinder(h=40,d=v2);
translate([-225,0,355])
cylinder(h=40,d=v2);
}
```

Shelling the Model geometry

1. Select the **union(){** loop.
2. Right-click and select **Copy**.
3. Click at the end of the script and press ENTER.
4. Right-click and select **Paste**.
5. Modify the values of the script, as shown

```
union(){
    loft (d1=v3,d2=v4);
mirror([1,0,0]) loft (d1=v3,d2=v4);
cylinder(h=40,d=v3);
translate([225,0,355])
cylinder(h=40,d=v4);
translate([-225,0,355])
cylinder(h=40,d=v4);
}
```

6. Enclose the two **union()** operations inside a **difference ()** operation, as shown.

```
difference (){
union(){
    loft (d1=v1,d2=v2);
mirror([1,0,0]) loft (d1=v1,d2=v2);
cylinder(h=40,d=v1);
translate([225,0,355])
cylinder(h=40,d=v2);
translate([-225,0,355])
cylinder(h=40,d=v2);
}
union(){
    loft (d1=v3,d2=v4);
mirror([1,0,0]) loft (d1=v3,d2=v4);
cylinder(h=40,d=v3);
translate([225,0,355])
cylinder(h=40,d=v4);
translate([-225,0,355])
cylinder(h=40,d=v4);
}
}
```

7. Click the **Render** icon to shell the model.

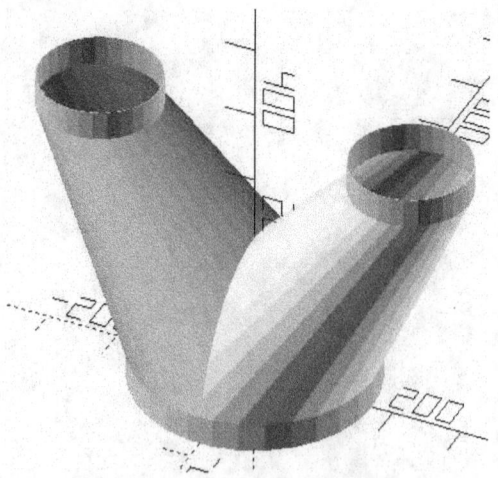

8. Click the **Save** icon on the **File** toolbar.
9. Type Tutorial_5 in the **File name** box and click **Save**.
10. Close the file.

Tutorial 6

In this example, you create the part shown below.

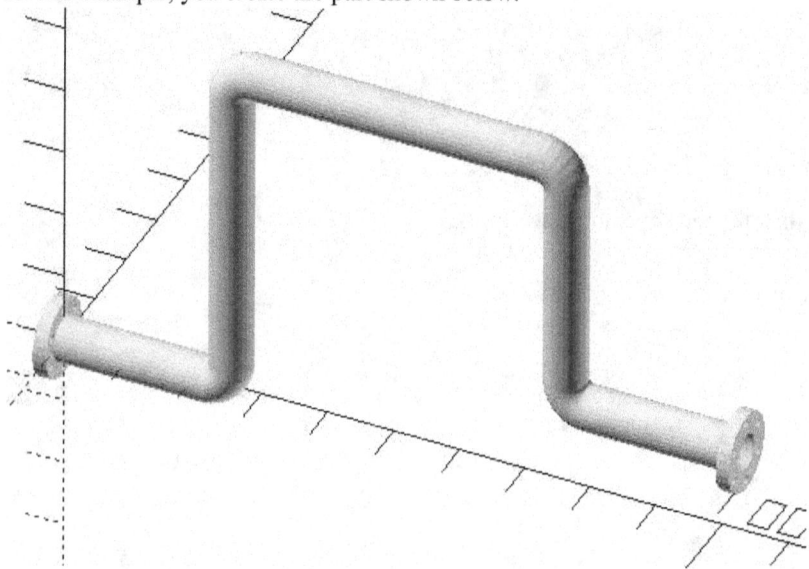

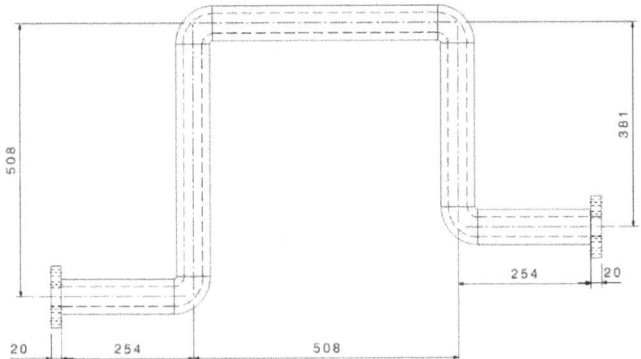

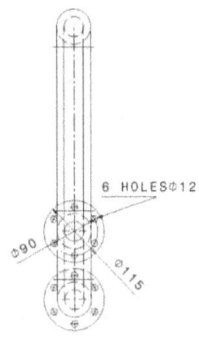

```
PIPE I.D. 51

PIPE O.D. 65
```

Creating a New document

1. Click the **Windows** icon located at the bottom left corner of the desktop. Next, click **O > OpenSCAD**.
2. Click the **New** button on the **Welcome to OpenSCAD** dialog.

Creating Pipes

1. Type the following two parameters.

d1=65;//outer diameter of the pipe
d2=51;//outer dimeter of the pipe
ir=5;//inner radius of the elbow
x=(d1/2+ir);//outer radius of the elbow
h1=254-x;//height of the first pipe
h2=508-2*x;//height of the second and third pipes
h3=381-2*x;//height of the fourth pipe
h4=h1;//height of the fifth pipe

2. Type the following script to create the **pipe** module, as shown.

```
module pipe(h,diameter1,diameter2) {
  difference(){
cylinder(h,d=diameter1,center=false);
cylinder(h,d=diameter2,center=false);
}
}
```

3. Type the following script to create a pipe and rotate it by 90 degrees about the Y axis, as shown.

```
rotate([0,90,0]){
pipe(h=h1,diameter1=d1,diameter2=d2);
}
```

4. Click the **Render** icon.

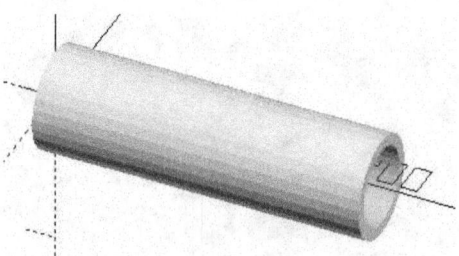

5. Type the following script to create a pipe and translate it in the X and Z directions, as shown.

    ```
    translate([h1+x,0,x]){
    pipe(h=h2,diameter1=d1,diameter2=d2);
    }
    ```

6. Click the **Render** icon.

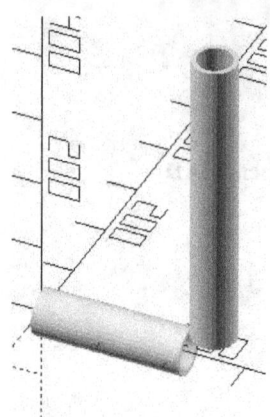

7. Type the following script to create a pipe and rotate and translate it, as shown.

```
translate([h1+2*x,0,h2+2*x])
rotate([0,90,0])
pipe(h=h2,diameter1=d1,diameter2=d2);
```

8. Click the **Render** icon.

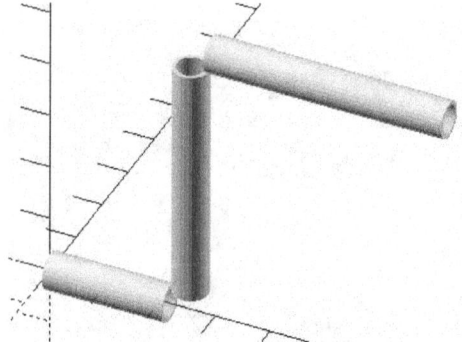

9. Type the following script to create a pipe and rotate and translate it, as shown.

```
translate([h1+h2+3*x,0,h2-h3+x]){
pipe(h=h3,diameter1=d1,diameter2=d2);
}
```

10. Click the **Render** icon.

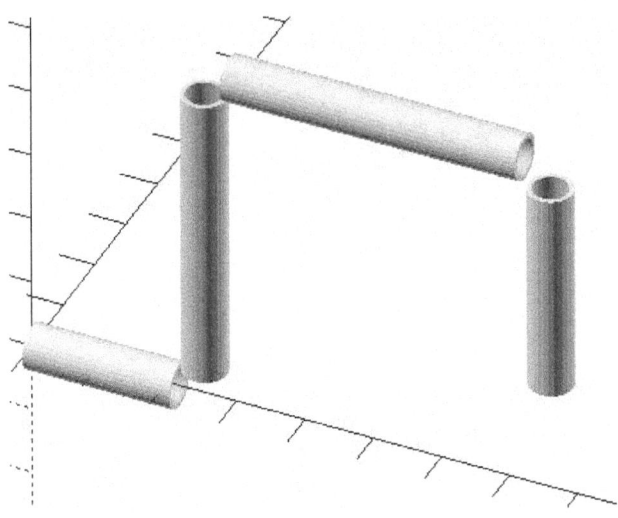

11. Type the following script to create a pipe and rotate and translate it.

```
translate([h1+h2+4*x,0,h2-h3])
rotate([0,90,0]){
pipe(h=h4,diameter1=d1,diameter2=d2);
}
```

12. Click the **Render** icon.

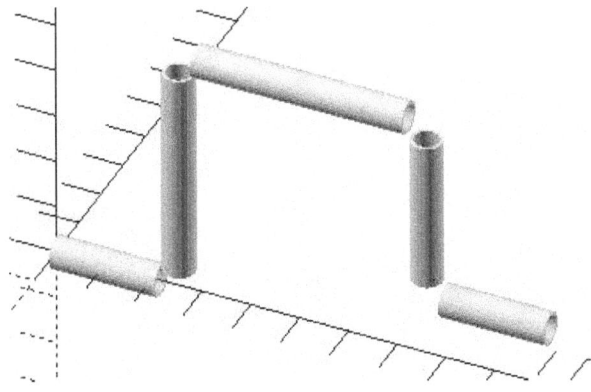

Creating Elbows

1. Type the following script to create the **elbow** module, as shown.

```
module elbow(a) {
   rotate_extrude(angle=a){
difference(){
translate([d1/2+5,0,0]) circle(d=d1);
translate([d1/2+5,0,0]) circle(d=d2);
}
}
}
```

2. Type the following script to create an elbow and rotate and translate it.

    translate([h1,0,x])
    rotate([-90,0,0]){
        elbow(a=90);
    }
3. Click the **Render** icon.

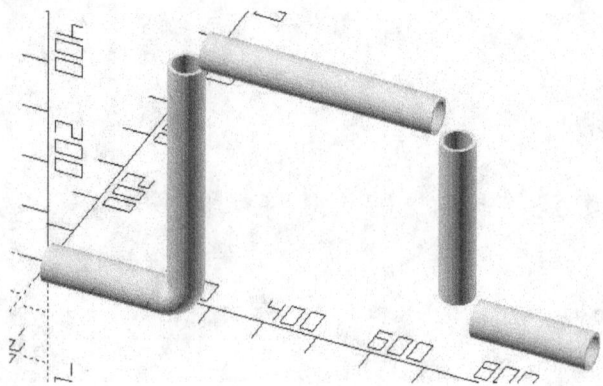

4. Type the following script to create an elbow and rotate and translate it.

translate([h1+d1+2*ir,0,h2+x])
rotate([90,-90,0]){
    elbow(a=90);
}
5. Click the **Render** icon.

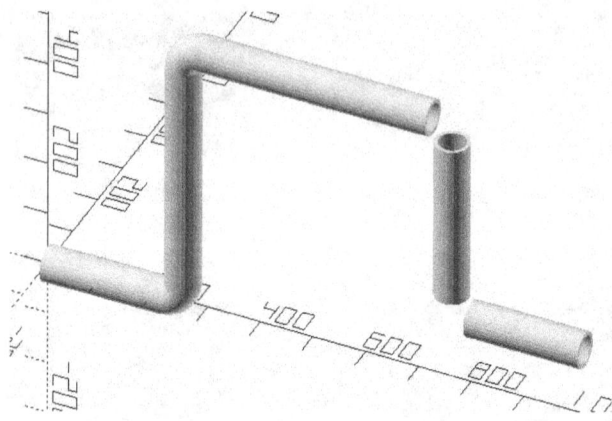

6. Type the following script to create an elbow and rotate and translate it.

    translate([h1+h2+2*x,0,h2+x])
    rotate([90,0,0]){
        elbow(a=90);
    }
7. Type the following script to create an elbow and rotate and translate it.

translate([h1+h2+4*x,0,h2-h3+x])

62

```
rotate([-90,90,0]){
   elbow(a=90);
}
```

8. Click the **Render** icon.

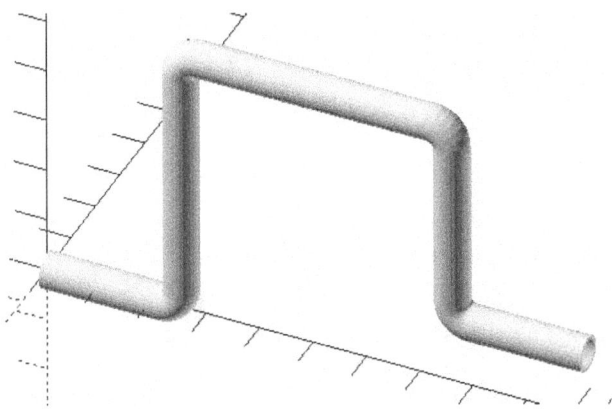

Creating the Flanges

1. Type the following script to create three cylinders and perform the **difference()** operation.

```
difference(){
cylinder(h=20,d=115,center=false);
 cylinder(h=20,d=51,center=false);
   translate([45,0,0])
    cylinder(h=20,d=12,center=false);
}
```

2. Click the **Render** icon.

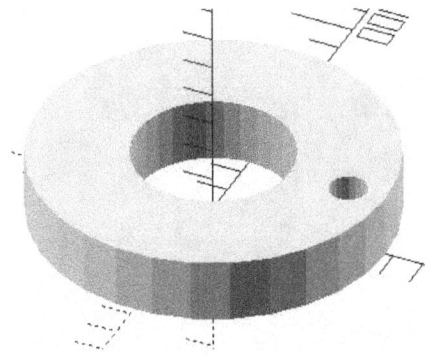

Next, you need to create a circular pattern of the hole.

3. Click at the end of the **cylinder(h=20,d=51,center=false);** function, and then press ENTER.
4. Type **for (z=[0:60:360]){** and press ENTER. It defines the **z** variable and its parameters.

z = variable
0 = start (initial position)
60 = increment
360 = end (end position)

5. Type **rotate([0,0,z])** and press ENTER.
6. Close the **for** loop.

```
difference(){
cylinder(h=20,d=115,center=false);
 cylinder(h=20,d=51,center=false);
   for (z=[0:60:360]){
       rotate([0,0,z])
    translate([45,0,0])
    cylinder(h=20,d=12,center=false);
 }
 }
```

7. Click the **Render** icon. The cylinder is patterned about the z axis.

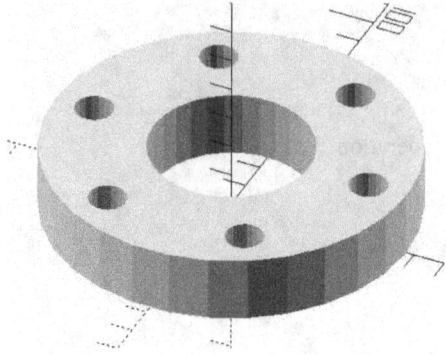

Next, you need to convert the flange into a module.

8. Click at the beginning of the **difference(){;** function, and then press ENTER.
9. Type the following script.
    module flange (i,t) {
10. Change the **for (z=[0:60:360]){** function to **for (z=[0:i:t]){**.
11. Close the **module** loop.

```
module flange (i,t) {
    difference(){
    cylinder(h=20,d=115,center=false);
     cylinder(h=20,d=51,center=false);
       for (z=[0:i:t]){
           rotate([0,0,z])
        translate([45,0,0])
        cylinder(h=20,d=12,center=false);
     }
     }
     }
```

12. Type the following script to create two flanges using the **flange** module.

```
//first_flange
translate([-20,0,0])
rotate([0,90,0])
flange (60,360);
```

64

```
//second_flange
translate([h1+h2+h4+4*x,0,h2-h3])
rotate([0,90,0])
flange (60,360);
```

13. Click the **Render** icon.

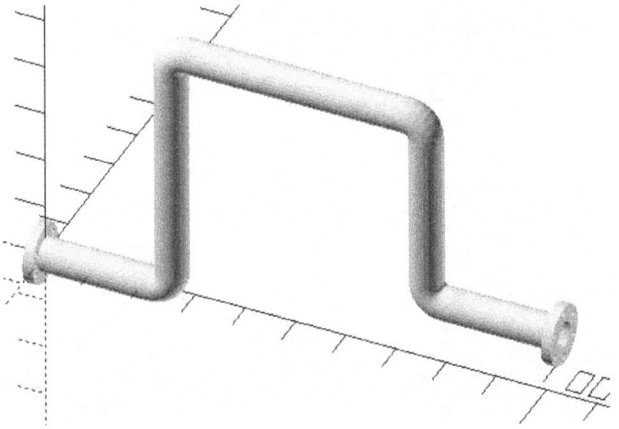

14. Save and close the file.

Tutorial 7

In this example, you create the part shown below and then modify it using the editing tools.

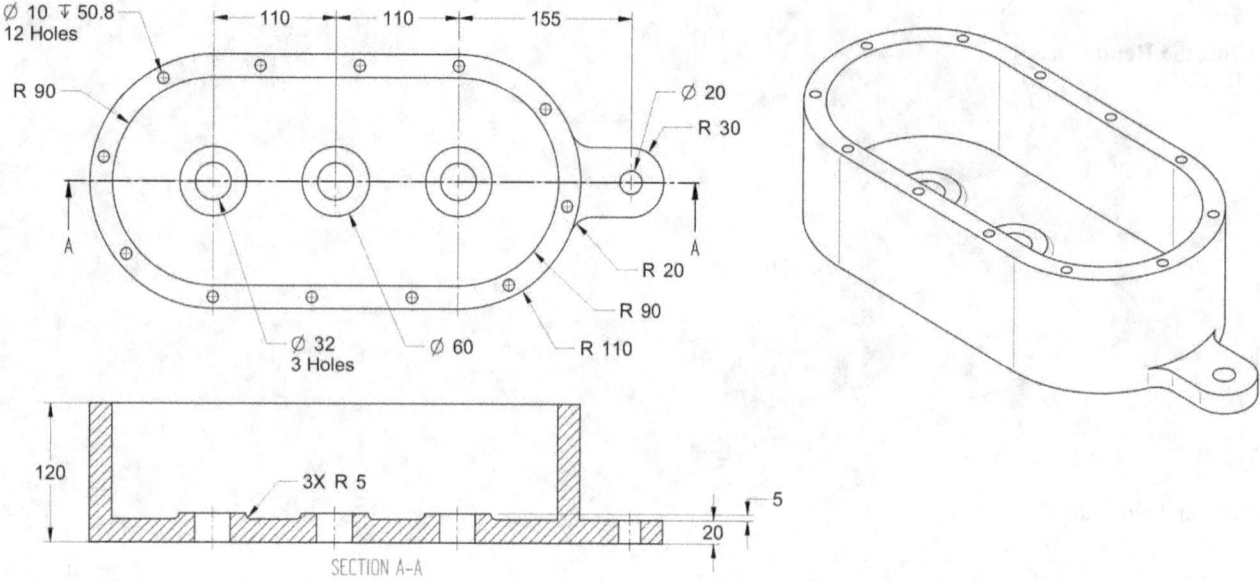

Creating a New document

1. Click the **Windows** icon located at the bottom left corner of the desktop. Next, click **O > OpenSCAD**.
2. Click the **New** button on the **Welcome to OpenSCAD** dialog.

Using the Hull function

1. Type the following parameters.

   $fn=125;
   d1=220;
   d2=180;
   d3=60;
   d4=20;
   d5=32;
   d6=10;
   h1=120;
   h2=100;
   h3=20;
   h4=5;
   h5=50.8;
   l1=220;
   l2=155;

2. Type the following script to create two extruded cylinders, as shown.

   cylinder(h=h1,d=d1);
   translate([220,0,0])

```
cylinder(h=h1,d=d1);
```

3. Click at the beginning of the first **cylinder()** function.
4. Type **hull(){**.
5. Close the loop at the end of the script.

```
hull(){
cylinder(h=h1,d=d1);
translate([220,0,0])
cylinder(h=h1,d=d1);
}
```

6. Click the **Render** icon.

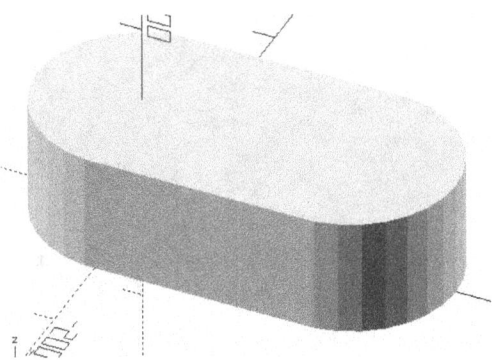

Using a Module

1. Click at the beginning of the **hull(){**function.
2. Type **module slot (height, diameter, length){**

slot = name of the module
height, diameter, and **length** are parameters of the module.

3. Press Enter.
4. Change the value of the **cylinder** and **translate** functions, as shown.
5. Close the loop at the end of the script.

```
module slot (height,diameter,length){
    hull(){
cylinder(h=height,d=diameter);
translate([length,0,0])
cylinder(h=height,d=diameter);
    }
}
```

Now, the module is defined, and you need to use the module to create features.

6. Type **slot (height=h1,diameter=d1,length=l1);**
7. Click the **Render** icon to create the slot.

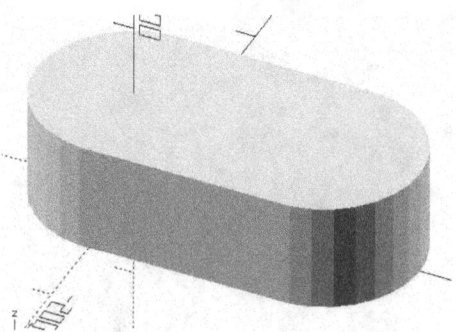

8. Type
   translate([l1,0,0])
   slot (height=h3,diameter=d3,length=l2);
9. Click the **Render** icon to create the slot.

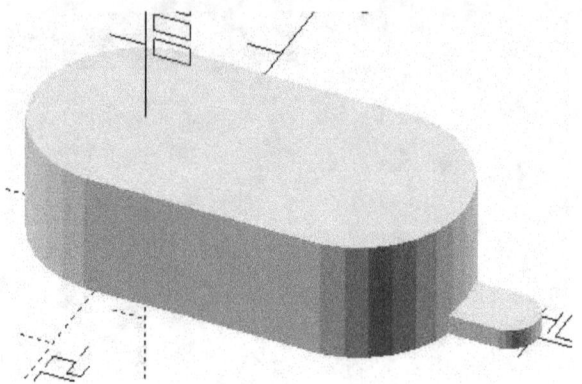

10. Create a **union ()** operation between the two slots.

    ```
    union(){
    slot (height=h1,diameter=d1,length=l1);
    translate([l1,0,0])
    slot (height=h3,diameter=d3,length=l2);
    }
    ```

11. Type
    translate([0,0,h3])
    slot (height=h2,diameter=d2,length=l1);
12. Create a **difference()** operation between the **union ()** and the new **slot**.
    ```
    difference(){
    union(){
    slot (height=h1,diameter=d1,length=l1);
    translate([l1,0,0])
    slot (height=h3,diameter=d3,length=l2);
    }
    translate([0,0,h3])
    slot (height=h2,diameter=d2,length=l1);
    }
    ```
13. Click the **Render** icon to create the slot.

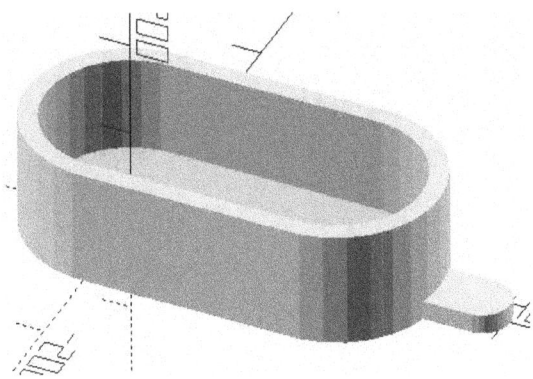

Creating cylinders with fillets

1. Type the following script to create a square and circle.

```
square(5);
translate([5,5,0])
circle(5);
```

2. Perform the **difference ()** operation between the square and the circle.

```
difference(){
    square(5);
translate([5,5,0])
circle(5);
}
```

3. Translate the sketch up to **d3/2** distance in the X-direction.
4. Use the **rotate_extrude()** function to revolve the sketch.

```
rotate_extrude(angle = 360) {
translate([d3/2,0,0])
difference(){
    square(5);
translate([5,5,0])
circle(5);
}
}
```

5. Add a cylinder to the revolved solid.

```
cylinder(h=h4,d=d3);
```

6. Create the **boss** module and change the values of the functions, as shown.

```
cylinder(h=          ,d=      );
    rotate_extrude(angle = 360) {
translate([          ,0,0])
difference(){
    square(5);
translate([5,5,0])
circle(5);
```

69

```
      }
     }
   }
```

7. Type the following script to create and translate the boss.

```
translate([0,0,h3])
 boss(h4,d3);
```

8. Click the **Render** icon to create the boss.

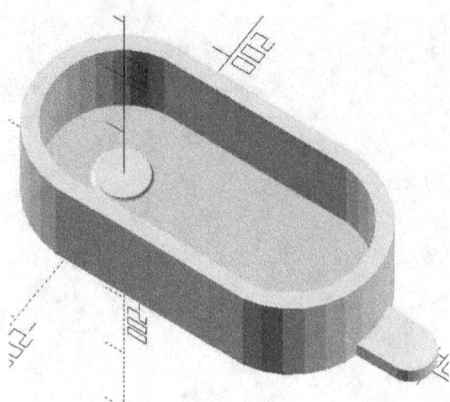

9. Enclose the **boss** module in the for loop, as shown.

```
for (x=[0:110:220]){
  translate([x,0,0])
    translate([0,0,h3])
     boss(h4,d3);
}
```

10. Click the **Render** icon to create the linear pattern of the boss.

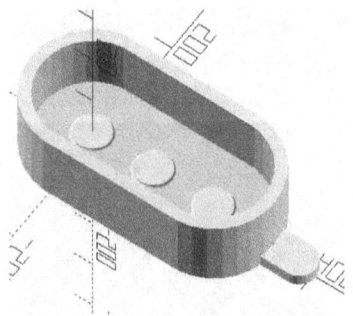

11. Select the **boss** module.
12. Right-click and select **Cut** from the shortcut menu.

```
27  slot (height=h2,diameter=d2,length=l1);
28  }
29  module boss(boss_height,boss_dia){
30  cylinder(h=boss_height,d=boss_dia);
31      rotate_extrude(angle             Undo
32  translate([boss_dia/2,0,0]            Redo
33  difference(){
34      square(5);                        Cut
35  translate([5,5,0])                    Copy
36  circle(5);                            Paste
37  }                                     Delete
38  }
39  }                                     Select All
40  for (x=[0:110:220]){
41      translate([x,0,0])
42  translate([0,0,h3])
43   boss(h4,d3);
44  }
45  for (x=[0:110:220]){
46      translate([x,0,0])
47      cylinder(h=h3+h4,d=d5);
48  }
```

13. Paste the **boss** module below the **slot** module.

```
13  module slot (height,diameter,length){
14      hull(){
15  cylinder(h=height,d=diameter);
16  translate([length,0,0])
17  cylinder(h=height,d=diameter);
18  }
19  }
20  ◀                Undo
21  differ           Redo
22  union (
23  slot (   Cut        diameter=d1,length=l1);
24  transl   Copy       0])
25  slo ➤    Paste      diameter=d3,length=l2);
26  }        Delete
27  transl   Select All 3])
28  slot (height=h2,diameter=d2,length=l1);
```

14. Create a **union ()** operation between the **difference ()** and **for** loop.

```
difference(){
union(){
slot (height=h1,diameter=d1,length=l1);
translate([l1,0,0])
slot (height=h3,diameter=d3,length=l2);
 }
translate([0,0,h3])
slot (height=h2,diameter=d2,length=l1);
}
for (x=[0:110:220]){
   translate([x,0,0])
translate([0,0,h3])
 boss(h4,d3);
}
```

15. Type the following script to create a linear pattern of the cylinders, as shown.

```
for (x=[0:110:220]){
   translate([x,0,0])
```

71

```
    cylinder(h=h3+h4,d=d5);
}
```

16. Create a **difference ()** operation between the **union ()** and **for** loop.

```
difference(){
    union(){
    difference(){
    union(){
    slot (height=h1,diameter=d1,length=l1);
    translate([l1,0,0])
    slot (height=h3,diameter=d3,length=l2);
     }
    translate([0,0,h3])
    slot (height=h2,diameter=d2,length=l1);
    }
    for (x=[0:110:220]){
       translate([x,0,0])
    translate([0,0,h3])
     boss(h4,d3);
    }
    }
    for (x=[0:110:220]){
       translate([x,0,0])
       cylinder(h=h3+h4,d=d5);
    }
```

17. Click the **Render** icon to create the linear pattern of the holes.

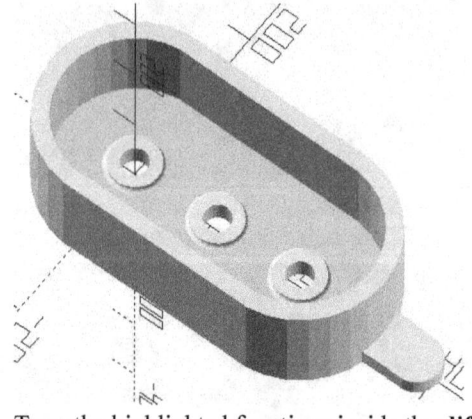

18. Type the highlighted functions inside the **difference()** operation, as shown.

```
difference(){
  union(){
  difference(){
  union(){
  slot (height=h1,diameter=d1,length=l1);
  translate([l1,0,0])
  slot (height=h3,diameter=d3,length=l2);
   }
  translate([0,0,h3])
```

```
slot (height=h2,diameter=d2,length=l1);
}
for (x=[0:110:220]){
  translate([x,0,0])
translate([0,0,h3])
 boss(h4,d3);
}
}
for (x=[0:110:220]){
  translate([x,0,0])
  cylinder(h=h3+h4,d=d5);
}

}
```

19. Click the **Render** icon to create the hole, as shown.

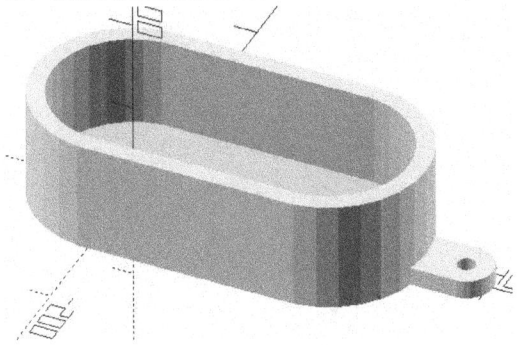

Creating the Fastening Holes

1. Create a **fastening_hole** module below the **boss** module.

```
module fastenning_hole(p1,p2,p3){
 translate([p1,p2,p3])
 cylinder(h=h5,d=d6);
 }
```

2. Create two **for** loops inside the **difference ()** operation, as shown.

```
difference(){
  union(){
  difference(){
  union(){
  slot (height=h1,diameter=d1,length=l1);
  translate([l1,0,0])
  slot (height=h3,diameter=d3,length=l2);
   }
  translate([0,0,h3])
  slot (height=h2,diameter=d2,length=l1);
  }
  for (x=[0:110:220]){
     translate([x,0,0])
  translate([0,0,h3])
```

```
   boss(h4,d3);
   }
 }
 for (x=[0:110:220]){
    translate([x,0,0])
    cylinder(h=h3+h4,d=d5);
 }
 translate([l2+l1,0,0])
 cylinder(h=h3,d=d4);
 for (x=[0:91.5:183]){
    translate([x,0,0])
 fastenning_hole(37,100,h1-h5);
 }
 for (x=[0:91.5:183]){
    translate([x,0,0])
 fastenning_hole(0,-100,h1-h5);
 }
}
```

3. Click the **Render** icon to create the linear pattern of the holes, as shown.

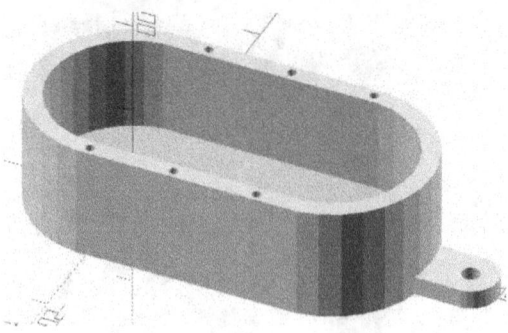

4. Create two **for** loops inside the **difference ()** operation, as shown.

```
for (z=[0:-50:-150]){
   rotate([0,0,z])
fastenning_hole(0,-100,h1-h5);
}
translate([220,0,0])
for (z=[0:-50:-150]){
   rotate([0,0,z])
fastenning_hole(0,100,h1-h5);
}
```

5. Click the **Render** icon to create the circular patterns of the holes, as shown.

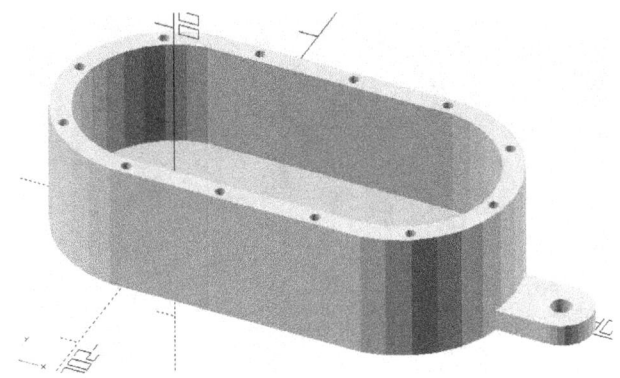

Creating the Fillets

1. Click the **New** icon on the **File toolbar**.
2. Create a square and two circles by entering the script, as shown.

```
translate([321.538,30,0])
 square([18.462,12.308]);
 translate([220,0,0])
 circle(110);
translate([340,50,0])
 circle(20);
```

3. Perform the **difference ()** operation between the square and the two circles, as shown.

```
difference(){
 translate([321.538,30,0])
 square([18.462,12.308]);
 translate([220,0,0])
 circle(110);
translate([340,50,0])
 circle(20);
 }
```

4. Click the **Render** icon.

5. Perform the **linear_extrude ()** operation to the resultant 2D object.

```
linear_extrude(height = 20){
 difference(){
 translate([321.538,30,0])
 square([18.462,12.308]);
 translate([220,0,0])
 circle(110);
 translate([340,50,0])
```

75

```
        circle(20);
        }
    }
```
6. Click the **Render** icon.

7. Convert the entire script into a module.

```
module fillet(f){
linear_extrude(height = 20){
difference(){
translate([321.538,30,0])
square([18.462,12.308]);
translate([220,0,0])
circle(110);
translate([340,50,0])
circle(f);
}
}
}
```

8. Select the entire script, right-click, and select **Cut**.
9. Switch to the main document and paste the copied script below the **fastening_hole** module, as shown.

```
32 - }
33  module fastenning_hole(p1,p2,p3){
34      translate([p1,p2,p3])
35      cylinder(h=h5,d=d6);
36  }
37  module fillet(f){
38  linear_extrude(height = 20){
39  difference(){
40      translate([321.538,30,0])
41      square([18.462,12.308]);
42      translate([220,0,0])
43      circle(110);
44  translate([340,50,0])
45      circle(f);
46  }
47  }
48  }
```

10. Type the following script at the bottom to create and mirror fillets.

```
fillet(20);
 mirror([0,1,0])
 fillet(20);
```

11. Click the **Render** icon.

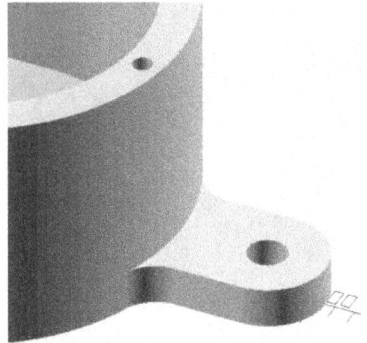

12. Save and close the file.